Letts

EDUCATIONAL

ADVANCED LEVEL

D0505014

Revise A2

Biology

Author

John Parker

Contents

Specification lists

AQA A Biology

MODULE	SPECIFICATION TOPIC	CHAPTER REFERENCE	STUDIED IN CLASS	REVISED	PRACTICE QUESTIONS
Unit 5 (M5)	Autotrophic nutrition	1.1			
	The biochemistry of photosynthesis	1.2			
	Respiration	1.3			
	Glycolysis and the Krebs cycle	1.3			
	Essential genetic terms	5.1			
	Meiosis	5.2			
	Sex linkage	5.3			
	Hardy–Weinberg Principle	5.3			
	Chi-squared: a statistical test	5.3			
	Classification of organisms	6.1			
	Investigations of ecosystems	7.1			
	Capture, mark, release, recapture	7.1			
	Water pollution	8.1			
	Pollution of the atmosphere	8.2			
Unit 6 (M6)	Digestion	2.2			
	Important structures in the small intestine	2.2			
	Digestion of food by ruminants	2.3			
	Lepidoptera; dietary changes through their life cycle	2.4			
	Structure and function neurones	3.1			
	The action potential	3.2			
	Synaptic transmission	3.2			
	Autonomic nervous system	3.4			
	The human eye	3.6			
	Homeostasis	4.1			
	The endocrine system	4.1			
	Temperature control in a mammal	4.2			
	Regulation of blood sugar level	4.3			
	Liver functions	4.3			
	The kidneys	4.4			
	The control of water balance	4.4			
	Adaptations to desert ecosystems	4.5			
	The behaviour of organisms	7.2			

Examination analysis

The A Level specification comprises 6 compulsory units, 3 AS and 3 A2. In unit tests 5, 6, and 8a, all questions are compulsory; they consist of structured questions and questions requiring extended answers.

Unit 5 1 hr 30 min examination 15%

Unit 6 1 hr 30 min examination 15%

Unit 8a 1 hr 45 min synoptic examination 10%

8b Centre-assessed coursework 10%

Note: the A2 centre-assessed coursework tests 8 different skills. Each may be assessed several times during the course. Marks awarded by your teacher are subject to change by external moderator. A key investigation is submitted as evidence of 7 of the skills. AS + A2 coursework counts towards the A Level.

AQA B Biology

MODULE	SPECIFICATION TOPIC	CHAPTER REFERENCE	STUDIED IN CLASS	REVISED	PRACTICE QUESTIONS
Unit 4 (M4)	The biochemistry of photosynthesis	1.2			
	Respiration	1.3			
	The action potential	3.2			
	The synapse	3.2			
	Functions of parts of the brain	3.4			
	Control of skeletal muscle	3.5			
	The mammalian eye	3.6			
	Homeostasis	4.1			
	Hormones	4.1			
	Temperature control	4.2			
	Regulation of blood sugar level	4.3			
	The kidneys	4.4			
	Essential genetic terms	5.1			
	Meiosis	5.2			
	Monohybrid and dihybrid inheritance	5.3			
	Classification of organisms	6.2			
	Natural selection	6.2			
	Speciation	6.2			
Unit 5 (M5)	Measurement in an ecosystem	7.1			
	Water pollution	8.1			
	Pollution of the atmosphere	8.2			
Unit 6 (M6)	Measurement in an ecosystem	7.1			
	Capture, mark, release, recapture	7.1			
Unit 7 (M7)	Microbial culture and measurement	9.2			
	Dilution plating	9.2			
	Large scale production	10.1			
	Biotechnology	10.1			
Unit 8 (M8)	Screening	5.4			
	Behaviour	7.2			
	Sign – stimulus release factors	7.2			

Examination analysis

The A Level specification comprises 6 units, 3 AS and 3 A2. The unit tests consist of structured questions and questions requiring extended answers. In A2, units 4 and 5 are compulsory and you can choose **one** unit from 6, 7 or 8. Each of units 6, 7 and 8 has compulsory questions plus a choice from two essay questions.

Unit 4   1 hr 30 min examination 15%
Unit 5a 1 hr 15 min examination 7.5% (including 3.5% synoptic questions)
Unit 5b Centre-assessed coursework 7.5% (including 2.5% synoptic questions)

Unit 6   2 hr examination 20% (including 14% synoptic questions) **or**
Unit 7   2 hr examination 20% (including 14% synoptic questions) **or**
Unit 8   2 hr examination 20% (including 14% synoptic questions)

Note: the A2 centre-assessed coursework tests 5 different skills. Each may be assessed several times during the course. Marks awarded by your teacher are subject to change by external moderator. The same investigation is submitted as evidence of the 5 skills. AS + A2 coursework counts towards the A Level.

OCR Biology

MODULE	SPECIFICATION TOPIC	CHAPTER REFERENCE	STUDIED IN CLASS	REVISED	PRACTICE QUESTIONS
Unit 4 (M4)	The biochemistry of photosynthesis	1.2			
	The biochemistry of respiration	1.3			
	The structure and functions of neurones	3.1			
	The action potential	3.2			
	Homeostasis	4.1			
	The endocrine system	4.1			
	The kidneys	4.4			
	Meiosis	5.2			
	Monohybrid and dihybrid inheritance	5.3			
	Classification and variation	6.1			
	Evolution	6.2			
	Investigation of ecosystems	7.1			
	Microbial culture and measurement	9.2			
Unit 5.01 (M5)	Plant growth regulators	3.7			
Unit 5.02 (M5)	Chi-squared: a statistical test	5.3			
	Applications of genetics	5.4			
Unit 5.03 (M5)	Classification and variation	6.1			
	Investigation of ecosystems	7.1			
	Pollution and effects	8.1, 8.2			
Unit 5.04 (M5)	Diversity of microorganisms	9.1			
	Microbial culture and measurements	9.2			
	Large scale production	10.1			
	Medical applications	10.2			
	Further gene transfer	10.3			
Unit 5.05 (M5)	Digestion and absorption	2.2			
	Ruminants and their microbial allies	2.3			
	Functions of parts of the brain	3.4			
	Control of skeletal muscle	3.5			
	The human eye and ear	3.6			
	Liver functions	4.3			

Examination analysis

The A Level specification comprises 6 units, 3 AS and 3 A2. The unit tests consist of structured questions and questions requiring extended answers. In A2 units 4 and 6.01 are compulsory. You can choose ONE option from 5.01–5.05, and choose either centre-assessed coursework or practical examination from 6.02 or 6.03.

2804 Unit 4	*1 hr 30 min examination 15%*			
2805 Unit 5.01	*1 hr 30 min examination 15%* **or**		**5.04**	*1 hr 30 min examination 15%* **or**
5.02	*1 hr 30 min examination 15%* **or**		**5.05**	*1 hr 30 min examination 15%* **or**
5.03	*1 hr 30 min examination 15%* **or**			
2806 Unit 6.01	*1 hr 15 min examination 10%*		**6.03**	*1 hr 30 min Practical examination 10%*
6.02	*Centre-assessed coursework 10%* **or**			

Note: *the A2 centre-assessed coursework tests 4 different skills. Each may be assessed several times during the course. Marks awarded by your teacher are subject to change by external moderator. A2 Level coursework is used to investigate the same skills as for AS Level but additional criteria must be satisfied. AS + A2 coursework counts towards the A Level. Alternatively a practical examination assesses the same practical skills.*

Edexcel Biology

MODULE	SPECIFICATION TOPIC	CHAPTER REFERENCE	STUDIED IN CLASS	REVISED	PRACTICE QUESTIONS
Unit 4 (M4)	Respiration	1.3			
	The structure and function of neurones	3.1			
	The action potential	3.2			
	The synapse	3.2			
	Homeostasis	4.1			
	Regulation of blood sugar level	4.2			
	The kidneys	4.4			
Unit 4A	Diversity of microorganisms	9.1			
	Gram positive and Gram negative	9.2			
	Microbial culture and measurement	9.2			
	Large scale production	10.1			
	Batch and continuous fermentation	10.1			
Unit 4B	Commercial production of beer	10.1			
Unit 4C	The synapse	3.2			
	Control of skeletal muscle	3.5			
Unit 5	The biochemistry of photosynthesis	1.2			
	Plant growth regulators	3.7			
	Essential genetic terms	5.1			
	Meiosis	5.2			
	Monohybrid and dihybrid inheritance	5.3			
	Applications of genetics	5.4			
	Classification	6.1			
	Evolution	6.2			
	Speciation	6.2			
	Investigation of ecosystems	7.1			
	Measurement in an ecosystem	7.1			
	Conservation	7.1			
	Further gene transfer	10.3			

Examination analysis

The A Level specification comprises 6 units, 3 AS and 3 A2. The unit tests consist of structured questions and questions requiring extended answers. In A2, units 4, 5 and 6 are compulsory. You can choose ONE option from A, B or C within unit 4. Unit 6 has a compulsory a synoptic examination, plus a choice of centre-assessed coursework or an alternative examination.

6104 Unit 4   1 hr 30 min examination 16.7% (inc. option A, B or C)

6105 Unit 5   1 hr 30 min examination 16.7%

6106 Unit 6   Centre-assessed coursework **or**
1 hr 20 min examination
1 hr 10 min examination (synoptic) 16.7% (inc. choice of **one** essay from two options)

WJEC Biology

MODULE	SPECIFICATION TOPIC	CHAPTER REFERENCE	STUDIED IN CLASS	REVISED	PRACTICE QUESTIONS
Unit 4 (M4)	The biochemistry of photosynthesis	1.2			
	The biochemistry of respiration	1.3			
	Glycolysis and the Krebs cycle	1.3			
	Digestion	2.2			
	Digestion in the duodenum and small intestine	2.2			
	Gram positive and Gram negative	9.2			
	Microbial culture and measurement	9.2			
	Dilution plating	9.2			
	Large scale production	10.1			
	Batch fermentation	10.1			
	Medical applications	10.2			
Unit 5 (M5)	Structure and function of the motor neurone	3.1			
	The action potential	3.2			
	Functions of parts of the brain	3.4			
	Control of skeletal muscle	3.5			
	The human ear	3.6			
	Homeostasis	4.1			
	The kidneys	4.4			
	Essential genetic terms	5.1			
	Monohybrid and dihybrid inheritance	5.3			
	Classification	6.1			
	Evolution	6.2			
	Artificial selection	6.2			
	Genetic conservation	6.2			
	Pollution	8.1, 8.2			
	Transgenic organisms	10.2			

Examination analysis

The A Level specification comprises 6 assessment units, 3 AS and 3 A2. The modular tests consist of structured questions and questions requiring extended answers. In A2, modules 4, 5 and 6 are compulsory.

Unit 4 *1 hr 40 min examination 15%*

Unit 5 *2 hr examination 20% (including synoptic questions)*

Unit 6 *Practical assessment by examination 15% (including synoptic)*

NICCEA Biology

MODULE	SPECIFICATION TOPIC	CHAPTER REFERENCE	STUDIED IN CLASS	REVISED	PRACTICE QUESTIONS
Module 4 (M4)	The biochemistry of photosynthesis	1.2			
	Respiration	1.3			
	Glycolysis and the Krebs cycle	1.3			
	The action potential	3.2			
	The synapse	3.2			
	Central nervous system	3.2			
	The human eye	3.6			
	Plant growth regulators	3.7			
	Phytochrome and the onset of flowering in plants	3.7			
	Homeostasis	4.1			
	The endocrine system	4.1			
	Temperature control in a mammal	4.2			
	Negative feedback	4.2			
	The kidneys	4.4			
	Hormone control of the kidneys; the role of ADH	4.4			
	Ecological conservation	7.1			
	Pollution	8.1, 8.2			
Module 5 (M5)	Essential genetic terms	5.1			
	Meiosis	5.2			
	Mendel and the laws of inheritance	5.3			
	Sex linkage	5.3			
	Hardy–Weinberg Principle	5.3			
	Applications of genetics	5.4			
	Classification of organisms	6.1			
	Continuous and discontinuous variation	6.1			
	Manipulation of reproduction	6.3			
	Artificial insemination	6.3			
	Superovulation and embryo transfer	6.3			

Examination analysis

The A Level specification comprises 6 compulsory modules, 3 AS and 3 A2. In modular tests 4, 5, and 6a, all questions are compulsory; they consist of structured questions and questions requiring extended answers.

Module 4   1 hr 30 min examination 16.7%
Module 5   1 hr 30 min examination 16.7%
Module 6a 1 hr examination (synoptic) 9.3%
Module 6b Centre-assessed coursework 7.4%

Note: the A2 centre-assessed coursework tests 8 different skill areas. Each is assessed in the context of ONE investigation during the A2 course. Marks awarded by your teacher are subject to change by external moderator. AS + A2 coursework counts towards the A Level.

AS/A2 Level Biology courses

All Biology GCE A Level courses currently studied are in two parts, with a number of separate units or modules in each part. Some units are further divided into sub-units and some have options to allow you to follow a specific path of interest. Some Examination Boards offer alternative externally marked practical examinations or the internal assessment of practical skills (subject to moderation).

In using this Revision Guide most students will have already completed the first half of the course, AS (Advanced Subsidiary). Some students will study both AS and A2 in one year. Some will go on to study the second part of the A Level course, called A2. Both groups of students are advised to use the Letts AS and A2 Revision Guides.

Advanced Subsidiary is assessed at the standard expected halfway through an A Level course, i.e. between GCSE and A Level. This means that AS and A2 courses are designed so that difficulty steadily increases:

- AS Biology builds from GCSE Science/ Biology
- A2 Biology builds from AS Biology.

IMPORTANT NOTE! Each Examination Board has included a common core of subject content in AS and in A2. Beyond the common core material the Examination Boards have included more varied content. Use the Examination Board Specification book or CD-ROM to identify subject content. The AS Biology Revision Guide includes material appropriate to A2 Level. Use the references to find A2 topics for your Examination Board which are to be found in the AS Biology Guide. You can be confident that these topics in the AS Guide will help you achieve understanding at A2 Level.

What are the differences between AS and A2?

There are three main differences:

(i)   A2 includes the more **demanding** concepts. (Understanding will be easier if you have completed the AS Biology course as a 'stepping stone'.)

(ii) There is a much greater emphasis on the skills of **application** and **analysis** than in AS. (Using knowledge and understanding acquired from AS is essential.)

(iii) A2 includes a substantial amount of **synoptic** material. (This is the drawing together of knowledge and skills across the modules of AS and A2. Synoptic investigative tasks and questions involving concepts across the specification are included.)

How will you be tested?

Assessment units

A2 Biology comprises three units or modules. The first two units are assessed by examinations.

The third component usually involves some method of practical assessment (this is dependent on the Examination Group). Examination Groups use **either** centre-assessed coursework **or** a practical examination.

Centre-based coursework involves practical skills marked by your teacher. The marks can be adjusted by moderators appointed by the awarding body. If a practical examination is an option, it is based on identical skills to the centre-assessed option.

Some groups also include another part to the third component. This is a short examination of further content.

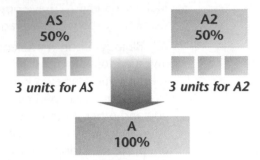

Tests are taken at two specific times of the year, January/February and June. It can be an advantage to you to take a unit test at the earlier optional time because you can re-sit the test, **(only once !)** The best mark from the two will be credited and the lower mark ignored.

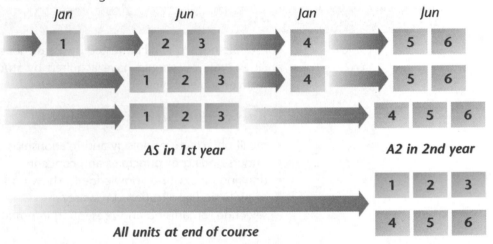

If you are disappointed with a module result, you can resit each module once. You will need to be very careful about when you take up a resit opportunity because you will have only one chance to improve your mark. The higher mark counts.

A unit or module can be retaken once but the complete course can be repeated in the future, should you wish to improve further.

A2 and synoptic assessment

Most students who study A2 have already studied to AS Level. There are three further units or modules to be studied. Some units are optional, so it is the choice of the Centre e.g. 'Applications of Genetics,' 'Environmental Biology,' or 'Microbiology and Biotechnology.'

Every A Level specification includes synoptic assessment at the end of A2. Synoptic questions draw on the ideas and concepts of earlier units, bringing them together in holistic contexts. Examiners will test your ability to inter-relate topics through the complete course from AS to A2. (See the synoptic chapter page 148).

Coursework

Coursework may form part of your work in Biology A2 course, depending on which specification you study. Where students have to undertake coursework it is usually for the assessment of practical skills.

Key skills

Your work in Biology AS and A2 can be used to gain a further award, the Key skills qualification. This helps you to develop important skills that are needed, whatever you do beyond A Level. The Key skills include: Application of number, Communication and Information Technology. There are three levels of award; Biology AS and A2 have opportunities to study one or more of the key skills, e.g. Communication – an average A2 student may be expected to achieve a Level 3.

A2 Biology is an excellent opportunity to achieve key skills. A coursework investigation can be used to achieve key skill levels. It is worth taking time to submit your coursework in a way which satisfies A2 Biology and key skill criteria. Discussing your experimental design, results and conclusions is beneficial to both you and other students. You must collect a portfolio evidence of each skill to show your level of competence. The awarding body specification will show opportunities of appropriate topics which can also be used to develop key skills. Additionally the QCA publication, *Introduction to key skills* will be helpful. Other subjects may be used to develop your key skills as well as AS and A2 Biology.

Key skills are in demand by Further Education institutions and by employers.

What skills will I need?

For A2 Biology, you will be tested by assessment objectives: these are the skills and abilities that you should have acquired by studying the course. The assessment objectives shown below.

Knowledge with understanding

- recall of facts, terminology and relationships
- understanding of principles and concepts
- drawing on existing knowledge to show understanding of the responsible use of biological applications in society
- selecting, organising and presenting information clearly and logically

Application of knowledge and understanding, and evaluation

- explaining and interpreting principles and concepts
- interpreting and translating, from one to another, data presented as continuous prose or in tables, diagrams and graphs
- carrying out relevant calculations
- applying knowledge and understanding to familiar and unfamiliar situations
- assessing the validity of biological information, experiments, inferences and statements

You must also present arguments and ideas clearly and logically, using specialist vocabulary where appropriate. Remember to balance your argument!

Experimental and investigative skills

Biology is a practical subject and part of the assessment of A2 Biology will test your practical skills. This may be done during your lessons or may be tested in a more formal practical examination. You will be tested on four main skills:

- planning
- implementing
- analysing evidence and drawing conclusions
- evaluating evidence and procedures.

The skills may be assessed in the context of separate practical exercises although more than one skill may be assessed in any one exercise. They may also be assessed all together in the context of a 'whole investigation'. An investigation may be set by your teacher or you may be able to pursue an investigation of your own choice.

You will receive guidance about how you practical skills will be assessed from your teacher. This Study Guide concentrates on preparing you for the written examinations testing the subject content of A2 Biology.

Different types of questions in A2 examinations

Questions in AS and A2 Biology are designed to assess a number of assessment objectives. For the written papers in AS Biology the main objectives being assessed are:

- recall of facts, terminology and inter-relationships
- understanding of principles and concepts and their social and technological applications and implications
- explanation and interpretation of principles and concepts
- interpreting information given as diagrams, photomicrographs, electron micrographs tables, data, graphs, passages
- application of knowledge and understanding to familiar and unfamiliar situations.

In order to assess these abilities and skills a number of different types of question are used.

In A2 Level Biology unit tests these include short answer questions, structured questions requiring both short answers and more extended answers, together with free-response and open-ended questions.

Short-answer questions

A question will normally begin with a brief amount of stimulus material. This may be in the form of a diagram, data or graph. A short-answer question may begin by testing recall. Usually this is followed up by questions which test understanding. Often you will be required to analyse data. Short answer questions normally have a space for your responses on the printed paper. The number of lines is a guide as to the amount of words you will need to answer the question. The number of marks indicated on the right side of the papers shows the number of marks you can score for each question part. Here are some examples. (The answers are shown in blue).

The diagram below shows a gastric pit.

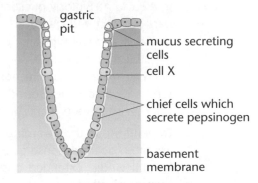

(a) (i) Label cell X   (1)
         oxyntic cell

    (ii) What is secreted by cell X?   (1)
         hydrochloric acid

(b) (i) Protein enters the stomach. What must take place before the
         hydrolysis of the protein begins?   (2)
         Hydrochloric acid acts on pepsinogen, to produce pepsin

    (ii) After the protein has been hydrolysed, what is produced?   (1)
         polypeptides

Structured questions

Structured questions are in several parts. The parts are usually about a common context and they often progress in difficulty as you work through each of the parts. They may start with simple recall, then test understanding of a familiar or unfamiliar situation. If the context seems unfamiliar the material will still be centred around concepts and skills from the Biology specification. (If a student can answer questions about unfamiliar situations then they display understanding rather than simple recall.)

The most difficult part of a structured question is usually at the end. Ascending in difficulty, a question allows a candidate to build in confidence. Right at the end technological and social applications of biological principles give a more demanding challenge. Most of the questions in this book are structured questions. This is the main type of question used in the assessment of both AS and A2 Biology.

The questions set at A2 Level are generally more difficult than those experienced at AS Level. A2 includes a number of higher level concepts, so can be expected to be more difficult. The key advice given by this author is:

- Give your answers in greater detail,

  Example: Why does blood glucose rise after a period without food?

  Answer: The hormone glucagon is produced X not enough for credit!)

  The hormone glucagon is produced which results in glycogen breakdown to glucose.

- Look out for questions with a 'sting in the tail'. A2 questions structured questions are less straight forward, so look for a 'twist'. This is identified in the example below.

When answering structured questions, do not feel that you have to complete a question before starting the next. Answering a part that you are sure of will build your confidence. If you run out of ideas go on to the next question. This will be more profitable than staying with a very difficult question which slows down progress, return at the end when you have more time.

Extended answers

In A2 and AS Biology questions requiring more extended answers will usually form part of structured questions. They will normally appear at the end of a structured question and will typically have a value of four to twenty marks. Longer questions are allocated more lines, so you can use this as a guide as to how many points you need to make in your response. Often for an answer worth ten marks the mark scheme would have around 12 → 14 creditable answers. You are awarded up to the maximum, ten marks, in this instance.

Depending on the awarding body, longer, extended questions may be set. These are often open response questions. These questions are worth up to twenty marks for full credit. Extended answers are used to allocate marks for the **quality of communication**.

Candidates are assessed on their ability to use a suitable style of writing, and organise relevant material, both logically and clearly. The use of specialist biological terms in context is also assessed. Spelling, punctuation and grammar are also taken into consideration. Here is a longer response question.

Question

Urea, glucose and water molecules enter the kidney via the renal artery. Explain what *can* happen to each of these substances.

In this question one mark is available for communication.   (Total 13 marks)

Urea, glucose and water molecules can pass through the blood capillaries in a glomerulus. ✓ This is as a result of ultrafiltration, ✓ as the podocytes of Bowman's capsule cause a pressure build up. ✓

The three substances pass down the proximal tubule. 100% glucose is reabsorbed in the proximal tubule ✓ so is returned to the blood. Carrier proteins on the microvilli aided by mitochondria, actively transport the glucose across the cells. ✓ Around 80% of the water is reabsorbed in the proximal tubule. ✓ Remaining water and urea molecules continue through the loop of Henlé. Urea continues through the distal tubule to the ureter then the bladder. ✓

More water can be reabsorbed with the help of the countercurrent multiplier. ✓ The ascending limb of the loop of Henlé ✓ actively transports Na^+ and Cl^- ions into the medulla. ✓ Water molecules leave the collecting duct by osmosis due to the ions in the medulla. ✓ Cells of the collecting duct are made more permeable to water by the hormone, ADH. ✓ Some water molecules pass into the capillary network and having been successfully reabsorbed ✓ Some water molecules continue down the ureters and into the bladder. ✓

Communication mark ✓

Remember that mark schemes for extended questions often exceed the question total, but you can only be awarded credit up to the maximum. In response to this question the candidate would be awarded the maximum of 13 marks which included one communication mark. The candidate gave two more creditable responses which were on the mark scheme, but had already scored a maximum. Try to give more detail in your answers to longer questions. This is the key to A2 success.

Exam technique

A2 builds from the skills and concepts acquired during the AS course. This Study Guide has been written in a similar style to the AS Biology Guide and incorporates many concepts. The Guide will help you cope as the A2 concepts ascend in difficulty. The chapters explain the ideas in small steps so that understanding takes place gradually. The final aim, of complete understanding of major topics is more likely.

Can I use my AS Biology Study Guide for A2?

YES! Some examination groups cover a topic at AS Level. A different examination group may cover the same topic at AS Level. Check out the A2 table for your specification. Every topic in the AS Guide is explained at a level suitable for A2 Level.

What are examiners looking for?

Whatever type of question you are answering, it is important to respond in a suitable way. Examiners use instructions to help you to decide the length and depth of your answer. The most common words used are given below, together with a brief description of what each word is asking for.

Define

This requires a formal statement. Some definitions are easy to recall.

Define the term transport.

This is the movement of molecules from where they are in lower concentration to where they are in higher concentration. The process requires energy.

Other definitions are more complex. Where you have problems it is helpful to give an example.

Define the term endemic.

This means that a disease is found regularly in a group of people, district or country. Use of an example clarifies the meaning. Indicating that malaria is invariably found everywhere in a country confirms understanding.

Explain

This requires a reason. The amount of detail needed is shown by the number of marks allocated.

Explain the difference between resolution and magnification.

Resolution is the ability to be able to distinguish between two points whereas magnification is the number of times an image is bigger than an object itself.

State

This requires a brief answer without any reason.

State one role of blood plasma in a mammal.

Transport of hormones to their target organs.

List

This requires a sequence of points with no explanation.

List the abiotic factors which can affect the rate of photosynthesis in pond weed.

carbon dioxide concentration; amount of light; temperature; pH of water

Describe

This requires a piece of prose which gives key points. Diagrams should be used where possible.

Describe the nervous control of heart rate.

The medulla oblongata ✔ *of the brain connects to the sino-atrial node in the right atrium, wall* ✔ *via the vagus nerve and the sympathetic nerve* ✔ *the sympathetic nerve speeds up the rate* ✔ *the vagus nerve slows it down.* ✔

Discuss

This requires points both for and against, together with a criticism of each point. (**Compare** is a similar command word).

Discuss the advantages and disadvantages of using systemic insecticides in agriculture.

Advantages are that the insecticides kill the pests which reduce yield ✔ *they enter the sap of the plants so insects which consume sap die* ✔ *the insecticide lasts longer than a contact insecticide, 2 weeks is not uncommon* ✔

Disadvantages are that insecticide may remain in the product and harm a consumer e.g. humans ✔ *it may destroy organisms other than the target* ✔ *no insecticide is 100% effective and develops resistant pests.* ✔

Suggest

This means that there is no single correct answer. Often you are given an unfamiliar situation to analyse. The examiners hope for logical deductions from the data given and that, usually, you apply your knowledge of biological concepts and principles.

The graph shows that the population of lynx decreased in 1980. Suggest reasons for this.

Weather conditions prevented plant growth ✔ *so the snowshoe hares could not get enough food and their population remained low* ✔ *so the lynx did not have enough hares (prey) to predate upon.* ✔ *The lynx could have had a disease which reduced numbers.* ✔

Calculate

This requires that you work out a numerical answer. Remember to give the units and to show your working, marks are usually available for a partially correct answer. If you work everything out in stages write down the sequence. Otherwise of you merely give the answer and if it is wrong, then the working marks are not available to you.

Calculate the Rf value of spot X. (X is 25 mm from start and solvent front is 100 mm)

$$Rf = \frac{distance\ moved\ by\ spot}{distance\ moved\ by\ the\ solvent\ front}$$

$$= \frac{25\ mm}{100\ mm} = 0.25$$

Outline

This requires that you give only the main points. The marks allocated will guide you on the number of points which you need to make.

Outline the use of restriction endonuclease in genetic engineering.

The enzyme is used to cut the DNA of the donor cell. ✔

It cuts the DNA up like this A T G C C G A T = A T   + G C C G A T ✔
                                T A C G G C T A   T A C G G C   T A

The DNA in a bacterial plasmid is cut with the same restriction endonuclease. ✔

The donor DNA will fit onto the sticky ends of the broken plasmid. ✔

If a question does not seem to make sense, you may have mis-read it. Read it again!

Some dos and don'ts

Dos

Do *answer the question*

No credit can be given for good Biology that is irrelevant to the question.

Do *use the mark allocation to guide how much you write*

Two marks are awarded for two valid points – writing more will rarely gain more credit and could mean wasted time or even contradicting earlier valid points.

Do *use diagrams, equations and tables in your responses*

Even in 'essay style' questions, these offer an excellent way of communicating biology.

Do *write legibly*

An examiner cannot give marks if the answer cannot be read.

Do *write using correct spelling and grammar. Structure longer essays carefully*

Marks are now awarded for the quality of your language in exams.

Don'ts

Don't *fill up any blank space on a paper*

In structured questions, the number of dotted lines should guide the length of your answer.

If you write too much, you waste time and may not finish the exam paper. You also risk contradicting yourself.

Don't *write out the question again*

This wastes time. The marks are for the answer!

Don't *contradict yourself*

The examiner cannot be expected to choose which answer is intended. You could lose a hard-earned mark.

Don't *spend too much time on a part that you find difficult*

You may not have enough time to complete the exam. You can always return to a difficult calculation if you have time at the end of the exam.

What grade do you want?

Everyone would like to improve their grades but you will only manage this with a lot of hard work and determination. You should have a fair idea of your natural ability and likely grade in Biology and the hints below offer advice on improving that grade.

For a Grade A

You will need to be a very good all-rounder.

- You must go into every exam knowing the work extremely well.
- You must be able to apply your knowledge to new, unfamiliar situations.
- You need to have practised many, many exam questions so that you are ready for the type of question that will appear.

The exams test all areas of the syllabus and any weaknesses in your Biology will be found out. There must be no holes in your knowledge and understanding. For a Grade A, you must be competent in all areas.

For a Grade C

You must have a reasonable grasp of Biology but you may have weaknesses in several areas and you will be unsure of some of the reasons for the Biology.

- Many Grade C candidates are just as good at answering questions as the Grade A students but holes and weaknesses often show up in just some topics.
- To improve, you will need to master your weaknesses and you must prepare thoroughly for the exam. You must become a better all-rounder.

For a Grade E

You cannot afford to miss the easy marks. Even if you find Biology difficult to understand and would be happy with a Grade E, there are plenty of questions in which you can gain marks.

- You must memorise all definitions.
- You must practise exam questions to give yourself confidence that you do know some Biology. In exams, answer the parts of questions that you know first. You must not waste time on the difficult parts. You can always go back to these later.
- The areas of Biology that you find most difficult are going to be hard to score on in exams. Even in the difficult questions, there are still marks to be gained. Show your working in calculations because credit is given for a sound method. You can always gain some marks if you get part of the way towards the solution.

What marks do you need?

The table below shows how your average mark is transferred into a grade.

average	80%	70%	60%	50%	40%
grade	A	B	C	D	E

Four steps to successful revision

Step 1: Understand

- Study the topic to be learned slowly. Make sure you understand the logic or important concepts.
- Mark up the text if necessary – underline, highlight and make notes.
- Re-read each paragraph slowly.

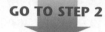

GO TO STEP 2

Step 2: Summarise

- Now make your own revision note summary:
  What is the main idea, theme or concept to be learned?
  What are the main points? How does the logic develop?
  Ask questions: Why? How? What next?
- Use bullet points, mind maps, patterned notes.
- Link ideas with mnemonics, mind maps, crazy stories.
- Note the title and date of the revision notes
  (e.g. Biology: Homeostasis, 3rd March).
- Organise your notes carefully and keep them in a file.

This is now in **short term memory**. You will forget 80% of it if you do not go to Step 3.
GO TO STEP 3, but first take a 10 minute break.

Step 3: Memorise

- Take 25 minute learning 'bites' with 5 minute breaks.
- After each 5 minute break test yourself:
  Cover the original revision note summary.
  Write down the main points.
  Speak out loud (record on tape).
  Tell someone else.
  Repeat many times.

The material is well on its way to **long term memory**.
You will forget 40% if you do not do step 4. **GO TO STEP 4**

Step 4: Track / Review

- Create a Revision Diary (one A4 page per day).
- Make a revision plan for the topic, e.g. 1 day later, 1 week later, 1 month later.
- Record your revision in your Revision Diary, e.g.
  Biology: Homeostasis, 3rd March 25 minutes
  Biology: Homeostasis, 5th March 15 minutes
  Biology: Homeostasis, 3rd April 15 minutes
  ... and then at monthly intervals.

Chapter 1

Energy for life

The following topics are covered in this chapter:

- *Autotrophic nutrition*
- *The biochemistry of photosynthesis*

- *Respiration*

1.1 Autotrophic nutrition

After studying this section you should be able to:

- *understand the principles of autotrophic nutrition*
- *relate the internal structure of a chloroplast to its function*

LEARNING SUMMARY

Different types of autotrophic nutrition

AQA A	M5
AQA B	M4, M5
EDEXCEL	M5
OCR	M4, M6
WJEC	M4
NICCEA	M4, M5

Key points from AS

- **The nitrogen cycle**
  Revise AS page 115
- **Photosynthesis**
  Revise AS page 109

Autotrophic nutrition is very important! **Autotrophic nutrition** means that simple inorganic substances are taken in and used to synthesise organic molecules. Energy is needed to achieve this. In **photo-autotrophic nutrition** light is the energy source. In most instances the light source is **solar energy**, the process being **photosynthesis**. Carbon dioxide and water are taken in by organisms and used to synthesise glucose, which can be broken down later during respiration to release the energy needed for life. By far the greatest energy supply to support food chains and webs is obtained from photo-autotrophic nutrition. Most producers use this nutritional method.

Chemo-autotrophic nutrition can also supply energy needs to some organisms. Simple inorganic substances are taken in and synthesised into organic molecules. **Chemical energy** is the source for this process.

Here are two examples of chemo-autotrophs:

Did you make the connection? Photo-synthesis is photo-autotrophic nutrition

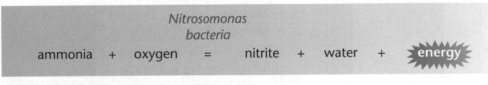

Nitrosomonas bacteria

ammonia + oxygen = nitrite + water + energy

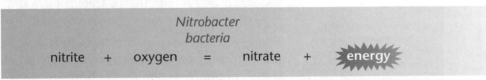

Nitrobacter bacteria

nitrite + oxygen = nitrate + energy

The energy released in each of the above reactions is the result of the oxidation of **inorganic** substances. During respiration it is organic chemicals which are oxidised.

The sulphur bacteria oxidise sulphur and release energy.

These bacteria are important in the nitrogen cycle.
The energy released in each reaction supplies the energy input for life for each of these bacteria.

Thiobacillus bacteria

sulphur + oxygen = sulphate + energy

Imagine these organisms in an underground cave, with the ability to support a complete food web without the need for light. Remarkable!

The chloroplast

AQA A	M5
AQA B	M4
EDEXCEL	M5
OCR	M4
WJEC	M4
NICCEA	M4

Chloroplasts are organelles in plant cells which photosynthesise. In a leaf they are strategically positioned to harvest the maximum amount of light energy. Most are located in the palisade mesophyll of leaves but they are also found in both spongy mesophyll and guard cells. There is a greater amount of light entering the upper surface of a leaf so the palisade tissues benefit from a greater chloroplast density.

The diagram below shows the structure of a chloroplast.

Remember that not all light reaching a leaf, may hit a chloroplast. Photons of light can be reflected or even absorbed by other parts of the cell. Around 4% of light entering an ecosystem is actually utilised in photosynthesis!

Even when light reaches the green leaf not all energy is fixed in the carbohydrate product. Just one quarter becomes chemical energy in carbohydrate.

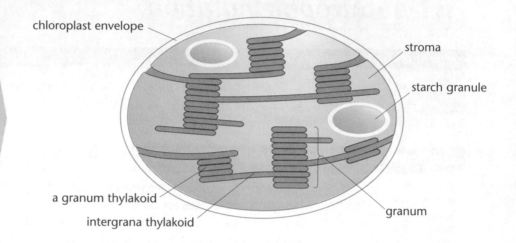

chloroplast envelope

stroma

starch granule

a granum thylakoid

intergrana thylakoid

granum

Structure and function

A system of **thylakoid membranes** is located throughout the chloroplast. These are flattened membranous vesicles which are surrounded by a liquid based matrix, the **stroma**.

Along the thylakoid membranes are key substances:

- chlorophyll molecules
- other pigments
- enzymes
- electron acceptor proteins.

Throughout the chloroplasts, circular thylakoid membranes stack on top of each other to form **grana**. Grana are linked by longer **intergrana thylakoids**. Grana thylakoids and intergrana thylakoids have different pigments and proteins. Each type has a different role in photosynthesis!

Do not be confused by the photosystems. They are groups of chemicals which harness light and pass on energy! Remember the information to understand the biochemistry of photosynthesis.

The key substances in the thylakoids occur in specific groups comprising of pigment, enzyme and electron acceptor proteins. There are two specific groups known as **photosystem I** and **photosystem II**.

The photosystems

Each photosystem contains a large number of chlorophyll molecules. As light energy is received at the chlorophyll, electrons from the chlorophyll are boosted to a higher level and energy is passed to pigment molecules known as the **reaction centre**.

> The reaction centre of photosystem I absorbs energy of wavelength 700 nanometres. The reaction centre of photosystem II absorbs energy of wavelength 680–690 nanometres. In this way light of different wavelengths can be harvested.

KEY POINT

1.2 The biochemistry of photosynthesis

After studying this section you should be able to:

- *recall and explain the biochemical processes of photosynthesis*
- *understand that glucose can be converted into a number of useful chemicals*
- *relate the properties of chlorophyll to the absorption and action spectra*
- *understand how the law of limiting factors is linked to productivity*

The process of photosynthesis

AQA A	M5
AQA B	M4
EDEXCEL	M5
OCR	M4, M6
WJEC	M4
NICCEA	M4

In examinations look out for parts of this diagram. There may be a few empty boxes where a key substance is missing. Will you be able to recall it?

- photosynthesis harnesses solar energy
- photosynthesis involves a light-dependent and light-independent reactions
- photosynthesis results in the flow of energy through an ecosystem.

The process of photosynthesis is summarised by the flow diagram below.

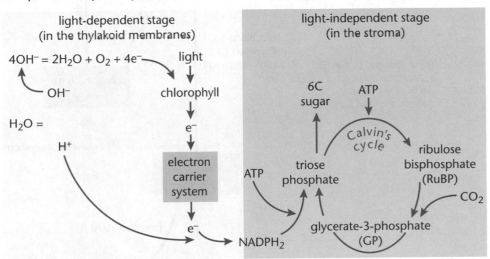

Light-dependent reaction

- Light energy results in the excitation of electrons in the **chlorophyll**.
- These electrons are passed along a series of electron acceptors in the thylakoid membranes, collectively known as the **electron carrier system**.
- Energy from excited electrons funds the production of **ATP** (adenosine triphosphate).
- The final electron acceptor forms **NADP$^+$**.
- Electron loss from chlorophyll causes the splitting of water (photolysis)

$$H_2O = H^+ + OH^- \quad \text{then} \quad 4OH^- = 2H_2O + O_2 + 4e^-$$

- Oxygen is produced, water to re-use, and electrons stream back to replace those lost in the chlorophyll.
- Hydrogen ions (H$^+$) from photolysis, together with NADP$^+$ form **NADPH$_2$**.

No ATP and NADPH$_2$ in a chloroplast would result in no glucose being made. Once supplies of ATP and NADPH$_2$ are exhausted then photosynthesis is ended. In exams look out for the 'lights out' questions where the light-independent reaction continues for a while until stores of ATP, NADPH$_2$ and GP are used up. These questions are likely to be graph based.

Light-independent reaction

- Two useful substances are produced by the light-dependent stage, ATP and **NADPH$_2$**. These are needed to drive the light-independent stage.
- They react with glycerate-3-phosphate (GP) to produce a triose sugar – **triose phosphate**.
- Triose phosphate is used *either* to produce a 6C sugar *or* to form **ribulose bisphosphate** (RuBP).
- The conversion of triose phosphate (3C) to RuBP begins Calvin's cycle and utilises ATP, which supplies the energy required.

Key points from AS

- **Gaseous exchange**
  Revise AS page 62
- **Photosynthesis**
  Revise AS pages 109–110.

- A RuBP molecule (5C) together with a carbon dioxide molecule (1C) forms two GP molecules (2 × 3C) to complete Calvin's cycle.
- The GP is then available to react with ATP and $NADPH_2$ to synthesise more triose sugar or RuBP.

How do the photosystems contribute to photosynthesis?

This can be explained in terms of the **Z scheme** shown below.

The **Z scheme**, so called because the paths of electrons shown in the diagram are in a 'Z' shape.

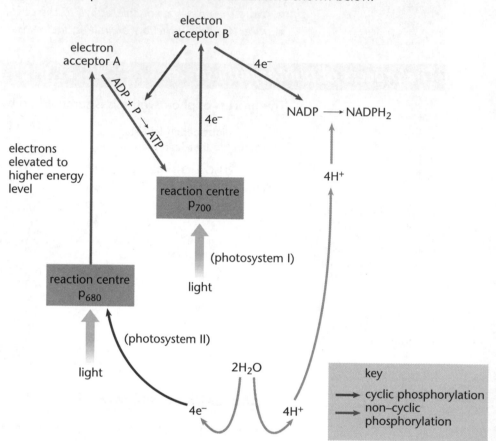

Non-cyclic photophosphorylation

- Light reaches the chlorophyll of both photosystems (P_{680} and P_{700}) which results in the excitation of electrons.
- Electron acceptors receive these electrons (**accepting** electrons is **reduction!**).
- P_{680} and P_{700} have become oxidised (**loss** of electrons is **oxidation!**).
- P_{680} receives electrons from the **lysis** (splitting) of water molecules and becomes neutral again.
- Lysis of water molecules releases oxygen which is given off.
- Electrons are elevated to a higher energy level by P_{680} to electron acceptor A and are passed along a series of electron carriers to P_{700}.
- Passage along the electron carrier system funds the production of ATP.
- The electrons pass along a further chain of electron carriers to NADP, which becomes reduced, and at the same time this combines with H^+ ions to form $NADPH_2$.

After analysing this information you will be aware that in cyclic photophosphorylation P_{700} donates electrons then some are recycled back, hence 'cyclic'. In non-cyclic photophosphorylation P_{680} electrons ultimately reach NADP never to return! Neutrality of the chlorophyll of P_{680} is achieved utilising electrons donated from the splitting of water. Different electron sources hence non-cyclic!

Cyclic photophosphorylation

- Electrons from acceptor B move along an electron carrier chain to P_{700}.
- Electron passage along the electron carrier system funds the production of ATP.

Photosynthetic pigments

> Chlorophyll *a* is the only photosynthetic pigment found in all green plants!

> The role of photosynthetic pigments is to absorb light energy.

Chlorophyll is not just one substance. There are several different chlorophylls, e.g. chlorophyll *a* and chlorophyll *b*.

- Each is a molecule which has a **hydrophilic head** and **hydrophobic tail**.
- The head always contains a **magnesium** ion and plays a key part in the absorbing or harvesting of light.
- The hydrophobic tail anchors to the thylakoid membrane.

As well as the chlorophylls there are other **accessory pigments**, e.g. carotenoids which also absorb light energy. There are a range of photosynthetic pigments found in different species.

The graphs below show the specific wavelengths of light which are absorbed by a range of pigments. The data for the **absorption spectrum** was collected by measuring the absorbance of a range of different wavelengths of light by a solution of each pigment, chlorophyll *a*, chlorophyll *b*, and carotenoids, **separately**. Following this, plants were illuminated at each wavelength of light, in turn, to investigate the amount of photosynthesis achieved at each wavelength. This data is shown in the **action spectrum**.

> The action spectrum shows the actual wavelengths which are used in photosynthesis.

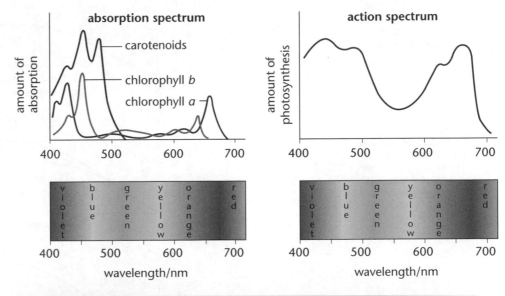

What can be learned from the graphs?

- Blue and red light are absorbed more, and so are key wavelengths for photosynthesis.
- Different pigments have different light absorptive properties.
- Groups of pigments in a chloroplast are therefore much better than just one as more energy can be harnessed for photosynthesis.
- The green part of the spectrum is not absorbed well; no wonder the plants look green as the light is reflected!

Which factors affect photosynthesis?

If any process is to take place then correct components and conditions are required. In the case of photosynthesis these are:

- light
- water
- carbon dioxide
- suitable temperature.

Additionally, it is most important that the chloroplasts have been able to develop their photosynthetic pigments in the thylakoid membranes. Without an adequate supply of magnesium and iron a plant suffers from **chlorosis** due to chlorophyll not developing. The leaf colour becomes yellow-green and photosynthesis is reduced.

Limiting factors

If a component is in low supply then productivity is prevented from reaching maximum. In photosynthesis **carbon dioxide** is a key limiting factor. The usual atmospheric level of carbon dioxide is 0.03%. In perfect conditions of water availability, light and temperature this low carbon dioxide level holds back the photosynthetic potential.

Clearly **light energy** is vital to the process of photosynthesis. It is severely limiting at times of partial light conditions, e.g. dawn or dusk.

Water is vital as a photosynthetic component. It is used in many other processes and has a lesser effect as a limiting factor of photosynthesis. In times of water shortage a plant suffers from a range of problems associated with other processes before a major effect is observed on photosynthesis.

A range of enzymes are involved in photosynthesis, therefore the process has an optimum **temperature** above and below which the rate reduces (so the temperature of the plant's environment can be limiting!).

Rate of photosynthesis is limited by light intensity from points A to B. After this a maximum rate is achieved – graph levels off.

Rate of photosynthesis limited by light intensity until each graph levels off. The 30°C graph shows that at 20°C temperature was also a limiting factor.

The lower level of CO_2 is also a limiting factor here. The fact that it holds back the process is shown by comparing both graph lines.

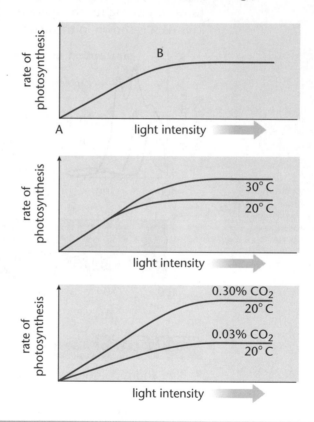

Compensation point

Photosynthesis utilises carbon dioxide whereas respiration results in its excretion. At night time during darkness a plant respires and gives out carbon dioxide. Photosynthesis only commences when light becomes available at dawn, if all other conditions are met. At one point the amount of carbon dioxide released by respiration is totally re-used in photosynthesis. This is the **compensation point**.

Another way of stating at compensation point is 'when the rate of respiration equals the rate of photosynthesis'.

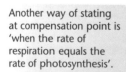

It is usual for a plant growing outside in warm conditions to have **two** compensation points every day.

Beyond this compensation point the plant may increasingly photosynthesise as conditions of temperature and light improve. The plant at this stage still respires producing carbon dioxide in its cells and all of this carbon dioxide is utilised. However, much more carbon dioxide is needed which diffuses in from the air.

In the evening when dusk arrives a point is reached when the rate of photosynthesis falls due to the decrease in light and the onset of darkness. The amount of carbon dioxide produced at one point is totally utilised in photosynthesis. Another compensation point has arrived!

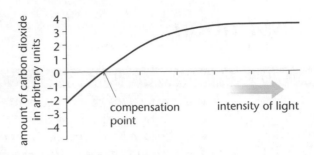

How useful is photosynthesis?

Without doubt it is a most important process because it supplies carbohydrates and gives off oxygen. There are many more benefits in that glucose is a 'starter' chemical for the synthesis of many other substances.

Cellulose, **amino acids**, and **lipids** are among the large number of chemicals which can be produced as a result of the initial process of photosynthesis.

The work of the Royal Mint produces the money to run the economy, photosynthesis supplies the **energy currency** for the living world.

The table shows some examples of where and how some carbohydrates are used.

> Many more substances are synthesised as a result of photosynthesis. Just a few are highlighted in this section.

carbohydrate	use
deoxyribose (monosaccharide)	DNA 'backbone'
glucose (monosaccharide)	leaves, nectar, blood as energy supply
sucrose (disaccharide)	sugar beet as energy store
lactose (disaccharide)	milk as energy supply
cellulose (disaccharide)	protective cover around all plant cells
starch (polysaccharide)	energy store in plant cells
glycogen (polysaccharide)	energy store in muscle and liver

Progress check

1 In a chloroplast where do the following take place:
   (a) light-dependent reaction
   (b) light-independent reaction?

2 (a) Which features do photosystems I and II share in a chloroplast?
   (b) Which photosystem is responsible for:
       (i) the elevation of electrons to their highest level
       (ii) acceptance of electrons from the lysis (splitting) of water?

3 (a) Complete the sentence by writing in the correct words.
       The compensation point of a plant is when the rate of ....
       equals the rate of ....

   (b) During a cloudless day in ideal conditions for photosynthesis, how many compensation points does a plant have? Give a reason for your answer.

4 List the three main factors which limit the rate of photosynthesis.

5 During the light-independent stage of photosynthesis which substances are needed to continue the production of RuBP? Underline the substances in your answer which are directly supplied from the light-dependent stage of photosynthesis.

1 (a) thylakoid membranes (b) stroma.

2 (a) Each photosystem contains a large number of chlorophyll molecules. Light energy is received at the chlorophyll where electrons are boosted to a higher level. Energy is passed to pigment molecules known as the **reaction centre**. The reaction centre of each photosystem absorbs energy (but of different wavelengths).
   (b) (i) photosystem II (ii) photosystem I.

3 (a) respiration; photosynthesis (b) Two. Around dawn and dusk there will come a time when the CO_2 produced as a result of respiration is totally used up in photosynthesis.

4 CO_2; light; temperature.

5 $NADPH_2$, ATP and CO_2.

1.3 Respiration

Adenosine triphosphate (ATP)

AQA A	M4
AQA B	M4
EDEXCEL	M4
OCR	M4
WJEC	M4
NICCEA	M4

> Remember that in photosynthesis ATP molecules are both synthesised then used to supply energy in the light-independent stage!

A glucose molecule has a high energy content. If all the energy was released at once then there would be severe temperature problems in a cell. It is important that energy liberation is in small bursts. This is achieved by using adenosine triphosphate (ATP) molecules. Substrates such as glucose are broken down in enzyme-catalysed stages to produce a number of ATP molecules.

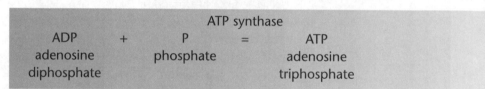

				ATP synthase		
ADP	+	P	=			ATP
adenosine diphosphate		phosphate				adenosine triphosphate

ATP is a molecule which is needed in all energy-requiring processes.

The ATP needs to be broken down to liberate its energy. This is done by an enzyme, **ATPase**.

> ATPase is a hydrolysing enzyme so that a water molecule is needed, but this is not normally shown in the equation.

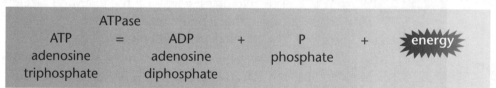

	ATPase						
ATP	=	ADP	+	P	+	energy
adenosine triphosphate		adenosine diphosphate		phosphate		

ATP is a **phosphorylated nucleotide**. Recall DNA structure which consists of nucleotides. Each nucleotide consists of an organic base, ribose sugar and phosphate group. ATP is a nucleotide with two extra phosphate groups! This is the reason for the term 'phosphorylated nucleotide'.

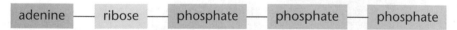

adenine	—	ribose	—	phosphate	—	phosphate	—	phosphate

> ATP is the energy currency of an organism.

The hydrolysis of the terminal phosphate group liberates the energy.

Uses of ATP

- muscle contraction
- active transport
- synthesis of macromolecules
- stimulates the breakdown of substrates to make even more ATP for other uses.

The biochemistry of respiration

Respiration is vital to the activities of every living cell. The flow diagram opposite shows stages in the breakdown of glucose and other substrates to produce a supply of ATP.

The two molecules of ATP are needed to begin the process. Each stage is catalysed by an enzyme, e.g. a decarboxylase removes CO_2 from a molecule.

ALERT! After the production of glycerate-3-phosphate the number of ATP molecules can be doubled. *Each* molecule of glycerate-3-phosphate gives rise to **20** molecules of ATP. Do not forget to take away the two ATPs at the start. So the total number of ATPs from one molecule of glucose is **38** (40 – 2). Count the ATPs in the diagram. Account for each ATP in the 38 total. This is a typical examination task!

The production of hydrogen atoms during the process can be monitored using DCPIP (dichlorophenol indophenol). It is a hydrogen acceptor and becomes colourless when fully reduced.

A mitochondrion

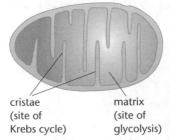

cristae
(site of
Krebs cycle)

matrix
(site of
glycolysis)

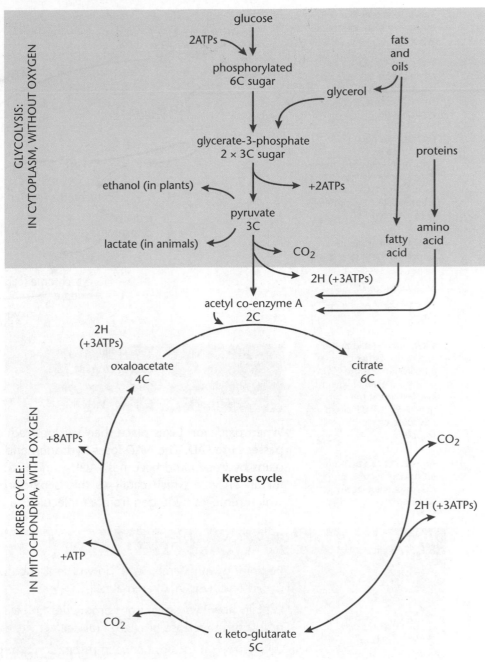

The flow diagram shows only the main stages of each process.

Glycolysis and the Krebs cycle

Both processes produce ATP from substrates but the Krebs cycle produces **many more** ATP molecules than glycolysis! Every stage in each process is catalysed by a specific enzyme. In aerobic respiration **both** glycolysis and the Krebs cycle are involved whereas in anaerobic respiration only glycolysis takes place.

The flow diagram shows that every time a stage produces two hydrogen atoms, in the presence of oxygen, three ATP molecules are produced. The role of these hydrogen atoms is shown in the **electron carrier system**.

Electron carrier system

The main feature of the electron carrier or electron transport system is that three ATPs are produced every time 2H atoms are transported. It takes place in the mitochondria.

This is sometimes known as the hydrogen carrier system.

The carrier, NAD, is nicotinamide adenine dinucleotide. Similarly, FAD is flavine adenine dinucleotide.

Hydrogen is not transferred to cytochrome. Instead, the 2H atoms ionise into $2H^+ + 2e^-$. H is passed via an intermediate co-enzyme Q to cytochrome.

Only the electrons are carried via the cytochromes.

e^- is an electron
H^+ is a hydrogen ion or proton.

Oxygen is needed at the end of the carrier chain as a hydrogen acceptor. This is why we need oxygen to live. Without it the generation of ATP along this route would be stopped.

An enzyme can be both an oxidoreductase and a dehydrogenase at the same time!

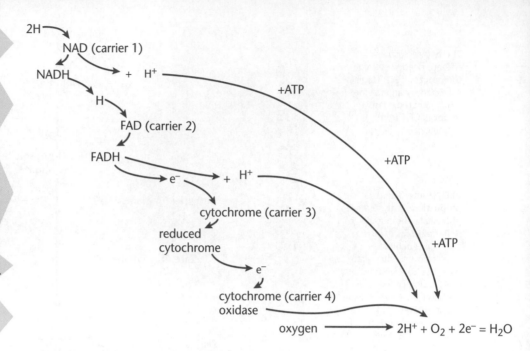

Oxidation	Reduction
Gain of oxygen	Loss of oxygen
Loss of hydrogen	Gain of hydrogen
Loss of electrons	Gain of electrons

When oxidation takes place then so does reduction, simultaneously, e.g. $NADH_2$ passes H to FAD. The NAD loses hydrogen and as a result becomes oxidised. FAD gains hydrogen and becomes $FADH_2$, and is therefore reduced. The generic term for an enzyme which catalyses this is **oxidoreductase**. Additionally an enzyme which removes hydrogen from a molecule is a **dehydrogenase**.

Progress check

1 Beginning with starch, write down the substances which could result from its use in glycolysis in an animal cell.

2 Explain how hydrogen atom production in cells during aerobic respiration results in the release of energy for cell activity.

3 Give **three** similarities between respiration and photosynthesis.

4 (a) Name the **four** carriers in the electron transport system in a mitochondrion. Give them in the correct sequence.

(b) Name the waste product which results from the final stage of the electron transport system.

5 For each of the following statements indicate whether a molecule would be oxidised or reduced.

(a) (i) loss of oxygen
   (ii) gain of hydrogen
   (iii) loss of electrons.

(b) Which type of enzyme enables hydrogen to be transferred from one molecule to another?

3 the stages of each process are catalysed by enzymes; both processes involve ATP; respiration involves GP in glycolysis and photosynthesis involves GP in the light-independent stage
5 (a) (i) reduced (ii) reduced (iii) oxidised.
   (b) Oxidoreductase

2 used in the electron transport system to produce ATP; 3 ATP molecules produced for every 2H atoms produced; ATP → ADP + P + energy released

4 (a) NAD → FAD → cytochrome → cytochrome oxidase
   (b) water

1 starch → glucose → glucose phosphate or phosphorylated glucose → glycerate-3-phosphate → pyruvate → lactate

Respiratory quotient

AQA A   M4
OCR   M4

It is sometimes useful to be able to deduce which substrate is being used in a person's metabolism at a specific time. This can be done if the volume of oxygen taken in, and the volume of carbon dioxide given out are measured. From this data the **respiratory quotient** (RQ) can be calculated.

$$RQ = \frac{\text{volume of carbon dioxide given off}}{\text{volume of oxygen taken in}}$$

The RQ values of the following substrates are well documented from previous investigations:

carbohydrate 1.0; protein 0.9; fat 0.7

It is interesting to know which substrate is being metabolised. It is necessary to view such data with caution. If a mixture of substrates is being used then the figure will be different from the above, e.g. an RQ of 0.8 could point to both protein and fat being used.

The graph below shows the different RQ values of a seed during different stages of germination.

Graph to show RQ values of barley through germination

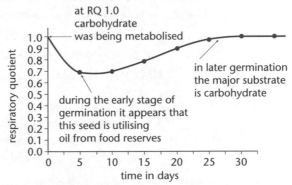

> If the volume of carbon dioxide given off is equal to the volume of oxygen taken in, what is the RQ value?

> Take great care in interpreting RQ data. This graph suggests that the seed begins with carbohydrate as a metabolite, changes to fat/oil then returns to mainly using carbohydrate. Any RQ which is not of the numbers given suggests a substrate combination is being used.

How is the RQ data collected?

The instrument called a **respirometer**, does this.

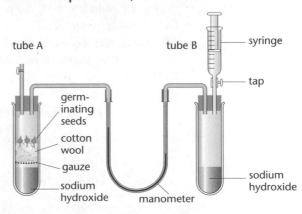

> Potassium hydroxide could be used instead of sodium hydroxide. They both absorb CO_2.

> If water replaces the sodium hydroxide then the carbon dioxide evolved can be measured.

- **Sodium hydroxide** absorbs all CO_2 from the air in the apparatus from the beginning.
- As the germinating seeds use oxygen and the pressure reduces in tube A so the manometer level nearest to the seeds rises.
- Any CO_2 excreted is absorbed by the sodium hydroxide solution.
- The syringe is used to return the manometer fluid levels to normal.
- The volume of oxygen used is calculated by measuring the volume of gas needed from the syringe to return the levels to the original values.

Sample question and model answer

Radioactivity is used to label molecules. They can then be tracked with a Geiger Müller counter.

In an experiment pondweed was immersed in water which was saturated with radioactive carbon dioxide ($^{14}CO_2$). It was illuminated for a time so that photosynthesis took place, the light was then switched off. The graph below shows the relative levels of some substances.

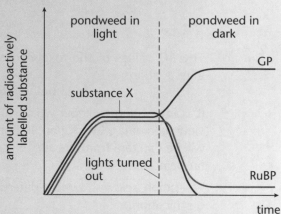

Always be ready to link the rise in one graph line with the dip of another. The relationship holds true here as substance X and RuBP are used up in the production of GP via the Calvin cycle. It is likely that some GP would have been used with substance X to make triose sugar. This is not shown on this graph.

Use the graph and your knowledge to answer the following questions.

(a) (i) Substance X is produced after a substance becomes reduced during the light-dependent stage of photosynthesis. Name substance X. [1]

NADPH$_2$, reduced nicotinamide adenine dinucleotide phosphate

(ii) Explain why substance X cannot be produced without light energy. [3]

- Light energy removes electrons from chlorophyll;
- the electrons are passed along the electron carrier chain;
- the electrons are needed to reduce NADP.

(b) Explain the levels of substance X, GP and RuBP after the lights were turned off. [6]

- It seems that substance X is used to make the other two substances because it becomes used up.
- Supply of substance X cannot be produced without light energy.
- GP is made from RuBP.
- GP levels out because more NADPH$_2$ is needed to make triose sugar or RuBP, the supply being exhausted.
- RuBP levels out at a low level because more NADPH$_2$ is needed to make GP.
- ATP is needed to make RuBP, ATP is needed to make GP.

ATP is not shown on the graph. Always be ready to consider substances involved in a process but not shown. Here it is worth a mark to remember that ATP is needed to continue the light-independent system of photosynthesis.

(c) After the lights were switched off glucose was found to decrease rapidly. Explain this decrease. [1]

- Glucose is used up in respiration to release energy for the cell.

(d) Give the specific sites of each of the following stages of photosynthesis in a chloroplast: [2]

(i) light-dependent stage   thylakoid membranes
(ii) light-independent stage.   stroma

Practice examination questions

1 The flow diagram below shows stages in the process of glycolysis.

2ATPs

glucose → phosphorylated → GP → substance X → lactate
6C sugar   6C sugar   glycerate-   3C
                            3-phosphate
                            (2 × 3C)

2ATPs

Use the information in the diagram and your knowledge to answer the questions below.

(a) Where in a cell does the above process take place? [1]

(b) Name substance X. [1]

(c) How many ATPs are *produced* during the above process? [1]

(d) Is the above process from an animal or plant?
Give a reason for your answer. [1]

(e) Under which condition could lactate be metabolised? [1]

[Total: 5]

2 The graph shows the relative amount of carbon dioxide taken in or evolved by a plant at different times during a day when the sun rose at 5.50 a.m.

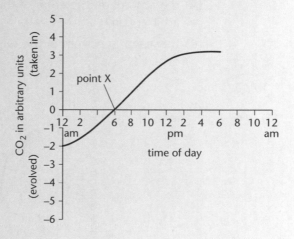

(a) Explain the significance of point X. [2]

(b) What name is given to point X? [1]

(c) Complete the graph between 6.00 p.m. and 12.00 a.m. [2]

[Total: 5]

3 An equation for the aerobic respiration of a lipid molecule included the production of 102 units of carbon dioxide.

The respiratory quotient was calculated as 0.7.

(a) Calculate the number (to nearest whole number) of oxygen units needed in the aerobic respiration of the lipid molecule. Show your working. [2]

(b) State the RQ value of:

(i) protein
(ii) carbohydrate. [2]

[Total: 4]

Practice examination questions (continued)

4 The diagram shows a respirometer set up to measure the amount of carbon dioxide produced by some insects.

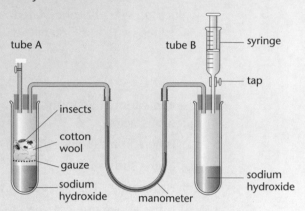

tube A   tube B — syringe
— tap
insects
cotton wool
gauze
sodium hydroxide
sodium hydroxide
manometer

Explain how the apparatus can be used to measure the O_2 taken in by the insects.   [4]

[Total: 4]

5 The chlorophyll in a pondweed consisted of several photosynthetic pigments. The graphs below show:

(A) the absorption spectrum of the pondweed's chlorophyll measured in arbitrary units

(B) the action spectrum of the same pondweed measured in cm^3 oxygen evolved.

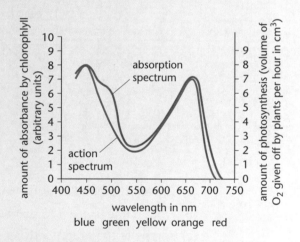

Use the graph and your knowledge to answer the questions.

(a) Explain the difference between the action and absorption spectra.   [2]

(b) Explain the effect of a wavelength of 525 nm on the rate of photosynthesis.   [1]

(c) How would the data for the action spectrum have been collected using the pondweed?   [1]

[Total: 4]

Practice examination questions (continued)

6 The flow diagram below shows part of the electron carrier system in an animal cell.

$$FADH \rightarrow FAD + H^+ + e^-$$

(a) Where in a cell does this process take place? [1]

(b) From which molecule did FAD receive H to become FADH? [1]

(c) Which molecule receives the electron produced by the breakdown of FADH? [1]

(d) As FADH becomes oxidised a useful substance is produced. Name the substance. [1]

[Total: 4]

7 The graph below shows the effect of increasing light intensity on the rate of photosynthesis of a plant where the concentration of carbon dioxide in the atmosphere was 0.03%.

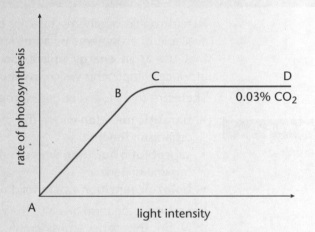

(a) Explain the effect of light intensity on the rate of photosynthesis between the following points on the graph:

(i) A and B

(ii) B and C

(iii) C and D. [3]

(b) Draw the shape of the graph which would result from a CO_2 concentration of 0.3%. [1]

[Total: 4]

Chapter 2

Nutrients

The following topics are covered in this chapter:

- *Heterotrophic nutrition*
- *Digestion*
- *Ruminants and their microbial allies*
- *Dietary changes in moths and butterflies*

2.1 Heterotrophic nutrition

After studying this section you should be able to:

- *understand the principles of heterotrophic nutrition and that it is divisible into holozoic, saprobiotic and parasitic nutrition*
- *understand the structure and function of dentition in herbivores and carnivores*

LEARNING SUMMARY

How do heterotrophic organisms feed?

AQA A	M6
EDEXCEL	M4
OCR	M5
WJEC	M4

Heterotrophic organisms take in **complex organic molecules**. These are made available in ecosystems by autotrophs. The heterotrophs obtain compounds which they use as an **energy source** and as **raw materials** to **build cell structures** and produce components which **assist cell processes**, e.g. assemble enzymes (proteins).

There are three forms of heterotrophic nutrition:

- **parasitic nutrition** where the organism obtains food from another *living* organism, the host
- **saprobiotic nutrition** where certain fungi and bacteria obtain food from dead organic material
- **holozoic nutrition** where solid or liquid food is taken into a 'gut' and digested.

Key points from AS

- **Saprophytic bacteria and fungi**
  Revise AS page 51

Ingestion is the term given to the way animals take in their food. The way in which they do this varies considerably throughout the biosphere. Organisms have adaptations which give them the ability to utilise food in their environment. Each organism has special features:

- **caterpillar** – powerful mandibles to shred leaves
- **butterfly** – coiled proboscis to insert into a nectary and suck up nectar
- **carnivore**, e.g. dog
  - long pointed canine teeth intersect to pierce the prey and prevent escape
  - large pointed molars to crush bone and cut through flesh
- **herbivore**, e.g. sheep
  - sharp incisors to cut plant material with horny pad to cut against
  - diastema (a space) to fit in plant material
  - flattened molars to grind plant material
  - sideways jaw movement to aid grinding.

The sheep is a ruminant. Find out more on p.42.

A caterpillar's head

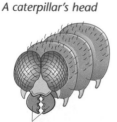

Mandibles – powerful jaw-like structures to shred plant material before consumption

A butterfly's head

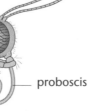

proboscis

A carnivore's head

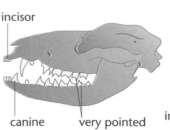

incisor

canine   very pointed molars

A herbivore's head

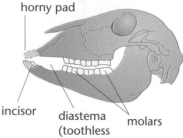

horny pad

incisor   diastema (toothless space)   molars

2.2 Digestion

After studying this section you should be able to:

- *understand the process of digestion in a range of organisms*
- *describe the roles of enzymes and hormones in the stages of digestion*
- *understand the absorption of nutrients in a mammalian gut*

How is food digested?

AQA A	M6
EDEXCEL	M4
OCR	M5
WJEC	M4

Key points from AS

- **Human digestive enzymes.**
  Revise AS page 51

Food consists mainly of insoluble molecules which must be broken down into smaller soluble ones to enable entry into the blood stream. In this way they can enter cells and be used by the organism. This is **assimilation**.

In the mouth the teeth **mechanically digest** food by breaking it up into smaller pieces. The **higher surface area** of these pieces enables the digestive enzymes to act efficiently. The greater the surface area of food molecules the greater the chance of these substrate molecules binding with the active sites of the digestive enzymes. The role of these enzymes is known as **chemical digestion**.

From the mouth to the anus the food passes through a tubular structure, the **alimentary canal**.

The diagram shows the arrangement of muscle tissue for the automatic movement of food through the alimentary canal. Smooth muscle cells lie in two different orientations, i.e. around the lumen (**circular**) or along the gut (**longitudinal**). The physical presence of food causes them to push from the oesophagus to the anus. This process transports gut contents and allows the stomach to churn food up to maximise enzyme contact with food substrate.

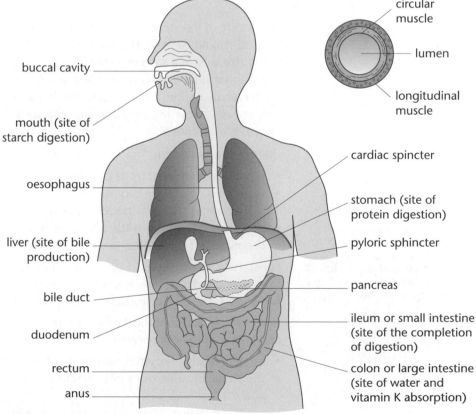

- circular muscle
- lumen
- longitudinal muscle
- buccal cavity
- mouth (site of starch digestion)
- oesophagus
- cardiac spincter
- stomach (site of protein digestion)
- pyloric sphincter
- liver (site of bile production)
- bile duct
- pancreas
- duodenum
- ileum or small intestine (site of the completion of digestion)
- rectum
- colon or large intestine (site of water and vitamin K absorption)
- anus

The passage of food through the alimentary canal requires assistance of both nervous and endocrine systems.

Digestion and the stomach

This is the site where protein digestion begins. The enzyme produced is pepsin and proteins are broken down into polypeptides.

pepsin
protein   →   polypeptides

Pepsin is an **endopeptidase** which breaks down the inner peptide links rather than the ends. In this way chains of amino acids (polypeptides) are formed.

Proteins are digested with the aid of cells in the **gastric glands** of the **stomach mucosa**.

Additionally babies secrete the enzyme rennin. This clots milk protein (caseinogen) so that it remains in the stomach long enough to be digested.

soluble → insoluble
caseinogen   casein

Section through stomach wall   *A gastric gland*

gastric pit

mucosa

sub-mucosa

circular muscle

longitudinal muscle

goblet cells to secrete mucus (the goblet invagination increases the surface area for secretions)

oxyntic cells to secrete hydrochloric acid

chief cells to secrete pepsinogen

basement membrane binding the cells of the gastric gland

The inside of the stomach wall is the **mucosa** where the important gastric pits are located. Lining each gastric pit is a lining of columnar epithelium. The cells of this lining tissue have a secretive role. Pepsin is produced in the stomach which creates a potential problem. Our cells are made of protein! Cells lining the gastric pit enable protein digestion without the danger of 'self' digestion. Mucus secreted by goblet cells across the stomach mucosa surface protects the stomach cell from its enzymes and from HCl.

Cells of the gastric pit:

- secrete **pepsinogen** from the **chief cells**
- secrete HCl from the **oxyntic** cells
- HCl activates the pepsinogen by hydrolysis
- **pepsin** is formed from the action of HCl on pepsinogen
- protein → polypeptides.

An inactive enzyme is a **precursor**. This is how the body prevents a potentially dangerous reaction taking place. The precursor is only activated when food is present! Look out for other precursors.

Peristalsis churns the food up for a period of around two to four hours. After this time the food will be ejected by peristaltic action via the pyloric sphincter. Sphincters are rings of muscle which hold substances in position for some time.

The partially digested food called chyme is at a low pH (around pH 2). This acidity can create problems because:

- other digestive enzymes do not work efficiently at low pH
- the small intestine does not secrete enough mucus to protect against harmful acid.

Again the body solves the problem by carefully controlled alkaline secretions which take effect in the first part of the small intestine, the **duodenum**.

Digestion in the duodenum and small intestine (ileum)

What a lot of enzymes! A number of enzymic secretions by the pancreas, duodenum and small intestine change the food molecules significantly. The table shows the major changes which take place in the duodenum and ileum.

PANCREATIC JUICE *Acts on food in duodenum and in ileum*	INTESTINAL JUICE *Acts on food in ileum*
enteropeptidase or enterokinase trypsinogen → trypsin	*amylase* starch → maltose
trypsin protein → polypeptides	*maltase* maltose → **glucose**
peptidase polypeptides → **amino acids**	*peptidase* polypeptides → **amino acids**
chymotrypsin chymotrypsinogen → chymotrypsin	*sucrase* sucrose → **glucose + fructose**
lipase lipids → **fatty acid + glycerol**	*nucleotidase* nucleotides → **pentose sugar + phosphate + organic bases**
nucleases nucleic acids → nucleotides	*lactase* lactose → **glucose + galactose**

Key – the substances in **bold** are small enough to be absorbed into the blood.

The enzymes above (in italics) require a pH around 7–7.5 to hydrolyse the food molecules efficiently.

Stomach acid; the problem solved!

HCl from the stomach entering the duodenum and ileum would quickly damage the duodenal and intestinal walls. Secretions are needed to neutralise the dangerous acid. These are supplied as:

- **alkaline** bile secreted by the liver, stored in the gall bladder and passed to the duodenum via the bile duct
- **alkaline** pancreatic juice which enters the duodenum at the same position.

Emulsification of lipids

Lipase alone could not break down lipids such as fats successfully unless lumps had been broken up into tiny droplets. This is **emulsification**.

> Increased surface area of food particles always aids enzyme action! Without increased surface area food would either become lodged in the alimentary canal or pass through to the anus with just the outside layer partially digested. This is another function of bile; high surface area of lipids.
>
> **KEY POINT**

When you revise for your exams put the enzymes into learning sequences. Begin with pepsin in the stomach then follow the break down of all proteins → polypeptides → amino acids. The final small soluble molecules are ideal for absorption.

Can you work out a sequence for carbohydrates?

Can you spot **two** precursors (inactive enzymes) in the table?

Do you remember pepsinogen in the stomach being activated by HCl? There is your clue.

Sometimes we produce so much stomach acid that it cannot be neutralised. Ulcers are the consequence! The drug-based remedy is a proton inhibitor to reduce oxyntic cell action.

Some people suffer agonising pain when bile cannot reach the duodenum (caused by perhaps gall stones in the bile duct). Accumulations of fats become lodged and stimulate pain receptors. A hospital visit is vital!

Important structures in the small intestine

The small intestine is the place where digestion is **completed** and the **soluble products** of digestion can be **absorbed**. The small intestine, around five metres long, gives the food a suitable period for the final digestive enzymes to be **secreted**, to **take effect** and for nutrients to be **absorbed**.

How do villi aid efficient digestion and absorption?

- Many **villi** provide a very high surface area for secretion of enzymes and absorption of nutrients.

- The individual outer layer of epithelium cells have **microvilli**, which give yet another boost to surface area.

- Near the base of villi the **crypts of Lieberkühn** are lined with enzyme secreting cells.

- Cells of the crypts of Lieberkühn continually divide mitotically and move to new higher positions up the villi.

- **Brunner's glands** secrete both mucus to lubricate the passing food and alkaline fluid to ensure neutralisation has been achieved.

- Movements of peristalsis continually push the food through the alimentary canal.

- Smooth muscle in the villi (**muscularis mucosa**) alternately contract and relax to ensure that a maximum rate of absorption is achieved. (Sluggish movement of the food would stagnate the absorptive processes.)

Try to make a list of all the different ways in which high surface areas are involved in the process of digestion.

Look at the goblet cells around the outer cell layer of each villus. They increase surface area still further!

Peristalsis only takes place when there is a sufficient quantity of food to stimulate receptors in the wall of the alimentary canal. Fibre in food keeps the muscle tone of the gut in prime condition.

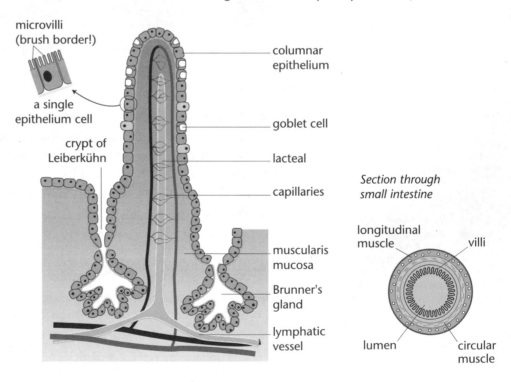

microvilli (brush border!)

a single epithelium cell

crypt of Leiberkühn

columnar epithelium

goblet cell

lacteal

capillaries

Section through small intestine

muscularis mucosa

Brunner's gland

lymphatic vessel

longitudinal muscle

villi

lumen

circular muscle

Remember that diffusion takes place down a concentration gradient. Movements of the gut and villi disturb the food contents and maximise the gradient! Concentration gradients between intestine lumen and blood are maximised.

Did you know?
Bacteria in the colon are able to synthesise vitamin K which passes into the blood.

Peristalsis continues the movement of digested food through the ileum. By the end, many of the nutrients (but not all!) will have been absorbed into the bloodstream. Substances such as amino acids and glucose pass into villus capillaries, reach the hepatic portal vein and subsequently enter a route to the body systems. Absorption by the villi is by **diffusion** and **active uptake**. Remaining in the intestinal contents are undigested substances such as fibre, water and some nutrients that were not able to contact villi. This mixture enters the colon or large intestine. Here much water is absorbed into the blood. The remains become more solid and are known as faeces. The faeces are stored in the rectum to be egested via the anus at a later stage.

What controls digestion?

AQA A	M6
OCR	M5

Digestion is coordinated by both the **nervous** and the **endocrine** systems working together. The tables below outline the roles of both systems.

Nervous control

digestive juice	stimulus and response
saliva	sight, thought, taste, smell of food causes an autonomic response via vagus nerve
gastric juice	stomach wall is stretched by food causes an autonomic response via vagus nerve
intestinal juice	contact of food with the small intestine wall stimulates Brunner's gland and crypts of Lieberkühn to make secretions

Note that the hormone enterogastrone '**switches off**' gastrin production. This is very useful since there is no food left in the stomach!

Hormonal control

stimulus	endocrine gland	hormone	effect
food in stomach	stomach mucosa	gastrin	causes oxyntic cells to secrete HCl
fat in ileum	intestinal mucosa	enterogastrone	inhibits gastrin production
food in ileum	intestinal mucosa	villikinin	stimulates the muscle in villi (muscularis mucosa) to contract rhythmically (which maximises nutrient absorption and empties lacteals into lymph vessels)
HCl in ileum	intestinal mucosa	secretin	stimulates the pancreas to secrete (i) alkaline ions (ii) watery fluid (but no enzymes!)
peptides and dipeptides in ileum	intestinal mucosa	pancreozymin (PZ)	stimulates the pancreas to produce enzymes
fats and oils	intestinal mucosa	cholecystokinin (CCK)	stimulates the gall bladder to contract

Note that secretin and pancreozymin both control the constituents of the pancreatic juice. Which controls the production of enzymes?

Have you ever had a fatty meal that 'lies heavily' on your stomach? It takes a long time to digest the fat because its digestion begins **after** the food has left the stomach. Fat is the last food component to leave. When fat reaches the ileum it stimulates CCK–PZ secretion, which results in bile being ejected from the gall bladder. Do you remember the functions of the bile? (p.39)

Take care with questions about hormones! The small intestine is very close to the pancreas but **all hormones** enter the general blood circulation. During digestion every blood vessel in the body will contain the hormones above but they only trigger responses to specific organs, e.g. pancreozymin passes through every vessel but only the pancreas responds by secreting enzymes.

Progress check

1 How does the body avoid damage to the duodenum by HCl from the stomach?
2 The diagram shows a gastric pit.
 (a) Name cells A and B.
 (b) In which part of the stomach wall are gastric pits located?
 (c) Explain how pepsinogen is activated in the stomach.
3 What is the effect of the hormone villikinin on digestion?

1 Alkaline bile secreted by the liver reaches the duodenum via the bile duct; alkaline pancreatic juice reaches the duodenum; neutralisation takes place; cholecystokinin (CCK) causes the gall bladder to empty; secretin stimulates the pancreas to secrete alkaline fluid.
2 (a) A mucus secreting cells, B chief (peptic) cells (b) (gastric) mucosa (c) food in stomach stimulates gastrin secretion; HCl secreted from the oxyntic cells; HCl activates the pepsinogen by hydrolysis; pepsin is formed.
3 Stimulates the muscle in villi (muscularis mucosa) to contract; villus movement maximises nutrient absorption; empties lacteals into lymph vessels; diffusion gradients between intestinal lumen and lumen maximised.

41

2.3 Ruminants and their microbial allies

After studying this section you should be able to:

- understand the roles of microorganisms in ruminants
- understand the association of partner organisms, ruminants and microorganisms as a mutualistic relationship

What is a ruminant?

A ruminant is a mammal which has a specialised digestive system in which the oesophagus leads to four sacs, the **rumen**, the **reticulum**, the **omasum** and the **abomasum**. In the rumen are **microorganisms** including bacteria and protoctistans. This relationship is an example of mutualism, where both microorganisms and the ruminant benefit from the association. Cows, sheep and antelopes are examples of ruminants.

How do ruminants digest their food?

- Plant material including grass is ingested prior to swallowing.
- Incisor teeth bite grass against the **diastema** or **horny pad** (together they act like a knife and cutting board in a kitchen!).
- Flat molars efficiently grind the plant material by a sideways jaw motion.
- Food boluses reach the **rumen** and **reticulum** where microorganisms commence their role.
- Bacteria produce **cellulase** enzymes which break down **cellulose** which surrounds every plant cell.
- After a time, partially digested boluses are **regurgitated** from the rumen to the mouth where further grinding takes place. (An even higher surface area to volume ratio enhances enzyme efficiency even more.)
- Complete breakdown of cellulose results in **volatile fatty acids** being produced and absorbed.
- Food enters the small intestine via the obomasum, the name for the true stomach, where the ruminant secretes its own enzymes.

Return of boluses to the mouth for further processing is 'chewing the cud!'

Try smelling the breath of a cow! Methane is excreted via the mouth (as well as the anus).

Both urea and protein are sources of nitrogen.

As some microorganisms die they are also digested to form a rich source of useful products including amino acids.

Digestion is never perfect. No wonder the flies have so much to feed on in faeces.

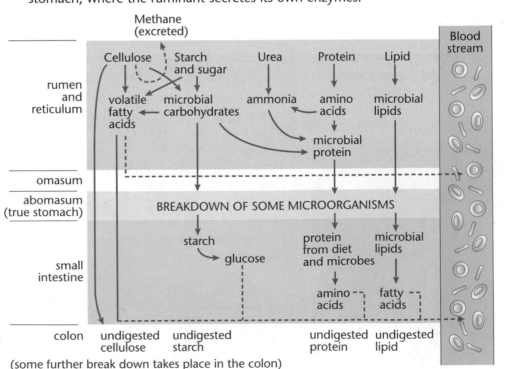

(some further break down takes place in the colon)

2.4 Dietary changes in moths and butterflies

After studying this section you should be able to:

- *understand the special adaptations of insects to their diet*

Dietary changes through the life cycle of moths and butterflies

AQA A   M6

Moths and butterflies are lepidopterous insects (e.g. cabbage white butterfly) and are characterised by a life cycle known as **complete metamorphosis**.

$$egg \rightarrow larva \rightarrow pupa \rightarrow adult\ (imago)$$

Larva (caterpillar) stage

- Is a specialist **feeding stage** when vast quantities of **protein-rich food** are **ingested**.
- Larvae radiate out **colonising new areas**, aiding the **dispersal** of the species.

Pupa (chrysalis) stage

- Signals a period of extreme **tissue re-organisation**.
- The rich protein supply of the larval stage gives rise to abundant amino acids which are assembled into different proteins, characteristic of the species.
- In time the **imago** or **adult** emerges from the pupal case.

Adult (imago) stage

- Is a specialist reproductive stage.
- Needs vast amounts of energy which are supplied by nectar, which the adult obtains by inserting its unwound proboscis into the nectary of a flower.
- Aids the dispersal of the species.

In an examination look out for a table like the one shown. You will have to put the ticks into a blank grid to show appropriate enzymes. Remember the main function of each stage then you will, logically, remember the correct enzymes. No digestive enzymes for the pupa. It does not feed!

Look out for unfamiliar data but, as always with examinations, apply the same principles learned during the course.

	STAGE IN LIFE CYCLE	
	larva	*adult*
food	leaves	nectar (sucrose)
feeding structures	mandibles	proboscis
amylase starch → maltose	✔	
maltase maltose → glucose	✔	
sucrase sucrose → glucose + fructose	✔	✔
protease protein → amino acids	✔	
lipase oils → fatty acids + glycerol	✔	

The table shows the different foods utilised by different stages of the same species. The flying action of the adults is fuelled by **sucrose** which only requires one enzyme, **sucrase**. The larva needs a complete range of enzymes.

Sample question and model answer

Be ready to relate the fall of one graph line to the rise of another!

The contents of an animal's stomach could be removed via an opening known as a fistula. The contents were measured after a meal.

The graph shows the relative levels of pepsinogen, substance X and pH.

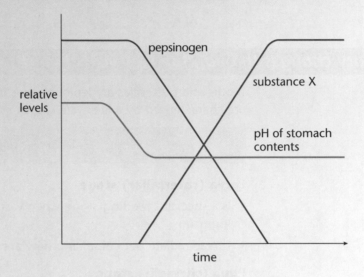

(a) Explain why the pH fell as a result of the meal entering the stomach.

- presence of food stimulates nervous system;
- presence of food in stomach stimulates gastrin production;
- as a result oxyntic cells produce more HCl;
- so the pH fell.

(b) (i) Suggest why the level of pepsinogen fell.

Pepsinogen is a precursor (inactive enzyme) which is activated by HCl in the stomach.

- HCl was used to change the pepsinogen into another substance;
- the fall in pepsinogen seems associated with increase in substance X.

(ii) Name substance X.

- pepsin

(c) The last food components to leave the stomach are fats and oils.

Explain how they are **efficiently** digested and absorbed.

This question may seem a formidable task. Do not worry! The mark scheme of an examiner may require only around 8 of these responses for full marks. This happens regularly on longer response questions.

- These lipids stimulate cholecystokinin to be secreted;
- gall bladder contracts to eject bile;
- bile emulsifies the lipids;
- pancreozymin stimulates pancreas to secrete lipase;
- secretin stimulates pancreas to secrete alkaline fluid
- (fluid environment essential for enzyme-controlled reactions);
- bile/alkali neutralises acid from stomach;
- lipase breaks down lipids into fatty acid + glycerol;
- absorption by villi which have a high surface area;
- absorption through microvilli;
- breakdown products pass into lacteals;
- lacteals emptied by regular contractions of the muscularis mucosa or smooth muscle of the villi.

Practice examination questions

1 (a) **Cellulose, starch, urea** and **protein** were eaten by a cow.

Complete the table by putting a tick in a box if the statement is true for that substance.

	Food substance			
	cellulose	starch	urea	protein
Can **only** provide a cow with nutrients because bacteria feed on it				
Results in the production of volatile fatty acids				
Contains a source of nitrogen				
Can be digested in the rumen and the small intestine				

[4]

(b) When a ruminant such as a cow chews grass, at first, it is swallowed and **partially** digested.

Describe how a cow makes sure that digestion is completed? [2]

(c) Bacteria live in the rumen of a cow.

Name this type of relationship. [1]

Describe **one** advantage to each partner. [2]

[Total: 9]

2 Complete the table by filling in the gaps.

stimulus	endocrine gland	hormone	effect
food in stomach	stomach mucosa	....	causes oxyntic cells to secrete HCl
fat in ileum	....	enterogastrone	inhibits gastrin production
food in ileum	....	villikinin	stimulates the muscle in villi (muscularis mucosa) to contract
.... in ileum	intestinal mucosa	....	stimulates the pancreas to secrete alkaline ions

[5]

[Total: 5]

Chapter 3

Control in animals and plants

The following topics are covered in this chapter:

- *Neurone: structure and function*
- *Nervous transmission*
- *The effects of drugs on synapses*
- *The central nervous system*

- *Control of skeletal muscle*
- *The human eye and ear*
- *Plant sensitivity*

3.1 Neurone: structure and function

LEARNING SUMMARY

After studying this section you should be able to:

- *describe the structure of a motor neurone and a sensory neurone*
- *understand the function of sensory, motor, bi-polar and multi-polar neurones*

The structure and functions of neurones

AQA A	M6
AQA B	M4
EDEXCEL	M4
OCR	M5
WJEC	M5
NICCEA	M4

Neurones are **nerve cells** which help to coordinate the activity of an organism by transmitting **electrical impulses**. In many organisms hormones contribute to this coordination (see next chapter).

KEY POINT

Important features of neurones.
1 Each has a **cell body** which contains a nucleus.
2 Each communicates via branched extensions called **dendrites**.
3 Some have long processes known as **axons**.
4 Neurones can transmit **electrical impulses**.

The nervous system consists of a range of different neurones which work in a network through the organs. The diagrams show two types of neurone.

> Note the direction of electrical impulses shown on the diagrams.

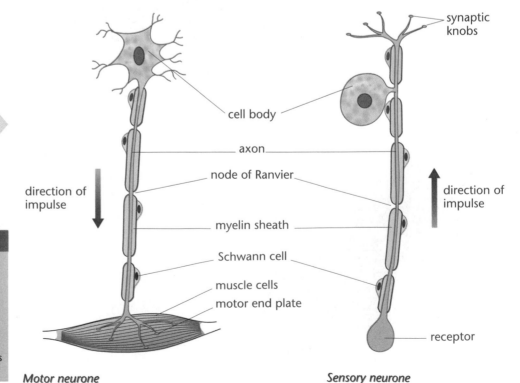

synaptic knobs

cell body

axon

node of Ranvier

direction of impulse

myelin sheath

Schwann cell

muscle cells

motor end plate

direction of impulse

receptor

Motor neurone

Sensory neurone

Key points from AS

- **The cell surface membrane**
  Revise AS pages 57–58
- **The movement of molecules in and out of cells**
  Revise AS pages 59–60
- **The specialisation of cells**
  Revise AS pages 41–42

Structure and function of the motor neurone

The motor neurone controls the contraction of muscle cells and has some important features.

- It has a **cell body** which includes the nucleus and many other cell organelles.
- Many **dendrites** radiate from the cell body to communicate with other neurones.
- A long process, the **axon**, containing many mitochondria, leads from the cell body.
- The axon can transmit a nerve impulse from the central nervous system for a considerable distance.
- The axon is insulated by the **myelin sheath**.
- The myelin sheath consists of a membrane of a **Schwann cell**, wrapped around the axon enclosing a layer of fat, giving the insulation property.
- At intervals there are gaps in the sheath where impulses are regenerated, each being known as a **node of Ranvier**.
- The axon terminates in a number of dendrites which contact muscle tissue via **motor end plates**.

A nervous impulse generated by a motor neurone usually results in the contraction of the muscle. The motor neurone is also known as an **effector** neurone. Some effector neurones control glands which secrete hormones when activated.

What about other neurones?

Multipolar neurones are found in the central nervous system (CNS). Their roles include memory, co-ordination and perception.

Bi-polar neurones link photosensitive cells in the retina to the optic nerve. They act as an intermediary between the two cells.

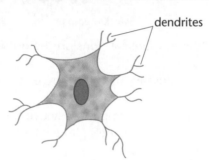
dendrites

Multipolar neurone

Bi-polar cell

What are the roles of the sensory neurones?

They have many similarities to the motor neurone, e.g. myelin sheath. However, there are important differences.

- The sensory neurones transmit impulses towards the central nervous system.
- Each sensory neurone has a receptor which responds to a specific stimulus, e.g. temperature change.
- The receptor responds to a local change by generating a nervous impulse which can be transmitted along the axon.
- The event at a receptor when a stimulus results in an impulse being generated is known as **transduction**.

Receptors

The more receptors there are in a position, the more sensitive it is, e.g. the fingers have many more touch receptors than the upper arm.

Did you know?
The umbilical cord has no receptors. It can be cut without any pain.

There are a range of different types of sensory cell around the body. Each type responds to different stimuli. Receptors are classified according to these different stimuli.

- **Photoreceptors**, respond to light, e.g. rods and cones in the retina.
- **Chemoreceptors**, respond to chemicals, e.g. taste buds on the tongue.
- **Thermoreceptors**, respond to temperature, e.g. skin thermoreceptors.
- **Mechanoreceptors**, respond to physical deformation, e.g. pressure receptors in the skin or hair cells in the ear.
- **Proprioreceptors**, respond to change in position in some organs, e.g. in muscles.

3.2 Nervous transmission

LEARNING SUMMARY

After studying this section you should be able to:

- *understand nervous transmission by action potential*
- *describe the mechanisms of synaptic transmission*

Transmission of an action potential along a neurone

AQA A	M6
AQA B	M4
EDEXCEL	M4
OCR	M5
WJEC	M5
NICCEA	M4

Neurones can 'transmit an electrical message' along an axon. However you must never write this in your answers. Instead of nerve impulse you must now use the term **action potential**.

The diagrams below show the sequence of events which take place along an axon as an action potential passes.

Resting potential

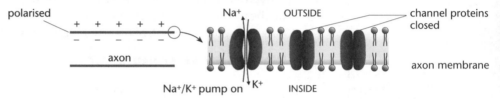

- There are 30 times more Na$^+$ ions on the outside of an axon during a resting potential.
- If any Na$^+$ ions diffuse in, then they are expelled by the '**sodium-potassium pump**'.
- The 'sodium-potassium pump' is an active transport mechanism by which a carrier protein, with ATP, expels Na$^+$ ions against a concentration gradient and allows K$^+$ ions into the axon.
- This creates a **polarisation**, i.e. there is a +ve charge on the outside of the membrane and a –ve charge on the inside.
- The **potential difference** can be measured at around –60 millivolts.

The Na$^+$/K$^+$ pump needs ATP to drive this activity. This is the reason for the large number of mitochondria in an axon.

Action potential

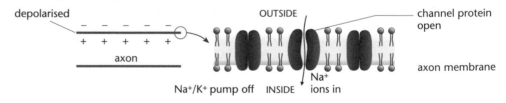

- As the action potential passes the **sodium-potassium pump is turned off**.
- Sodium channel proteins open to allow Na$^+$ ions into the axon.
- There is now a –ve charge on the outside and a +ve charge on the inside known as **depolarisation**.
- The potential difference changes to around +60 millivolts.
- The profile of the action potential, shown by an oscilloscope, is always the same.

Refractory period

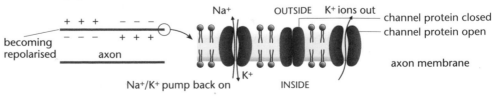

- A K$^+$ channel opens so K$^+$ ions leave the axon.
- The resting potential is regained by the 'sodium-potassium pump' being activated again.

> Only when the resting potential has been achieved is the axon ready to allow the passage of another action potential!

Measuring an action potential

- The speed and profile of action potentials can be measured with the help of an oscilloscope and glass electrodes.
- The profile of the action potential for an organism always shows the same pattern, like the one shown.
- The changes in potential difference are tracked via a time base.
- Using the time base you can work out the speed at which action potentials pass along an axon as well as how long one lasts.

The diagram shows the arrangement of the apparatus and the enlarged diagram shows the typical profile of an action potential.

If action potentials are the same throughout the nervous system, how can we differentiate different stimuli?

Answer – action potentials are propagated in specific patterns along an axon, e.g. like this. The action potentials are in volleys.

1111 1111 1111 1111

Each 1 represents an action potential on an oscilloscope screen.

Much of the data about action potentials was collected using neurones from the squid. The reason for this is that this organism has giant axons, suitable for the insertion of glass electrodes. The action potential profile can be applied to transmission along human neurones.

*The movement of an action potential along an axon is also known as **saltatory conduction**. This is due to the role of Na$^+$ ions. The impulse is regenerated at every node of Ranvier.*

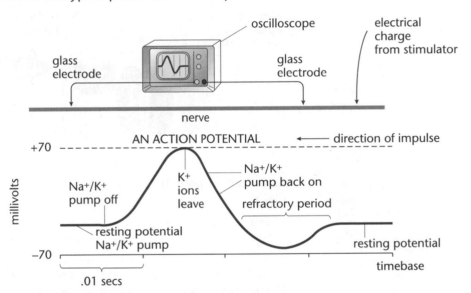

- The front of the action potential is marked by the Na$^+$/K$^+$ pump being off.
- The potential difference increases to around +70 millivolts as the Na$^+$ ions stream into the axon.
- K$^+$ ions leave the axon.
- The Na$^+$/K$^+$ pump re-starts and ultimately polarisation is re-established by the end of the refractory period.
- During the refractory period no other action potential can pass along the axon, which makes each action potential separate or discrete.

Progress check

What is the function of each of the following?

(a) receptor
(b) axon
(c) myelin sheath
(d) terminal dendrites.

(d) terminal dendrites have motor end plates which can stimulate muscle tissue to contract.
(c) myelin sheath is a membrane enclosing fat which acts as an insulator
(b) transmit action potential with the help of mitochondria
(a) receptors respond to stimulus by producing an action potential

How do neurones communicate with each other?

The key to links between neurones are structures known as **synapses**. Terminal dendrites branch out from neurones and terminate in **synaptic knobs**. The diagram below shows a synaptic knob separated from an interlinking neurone by a synapse.

Remember an impulse can 'cross' a synapse by chemical means and the route is in ONE direction only. They cannot go back!

A synapse which conducts using acetylcholine is known as a cholinergic synapse.

There are **two** types of synapse:
- **excitatory** which can stimulate an action potential in a linked neurone
- **inhibitory** which can prevent an action potential being generated.

terminal dendrite
mitochondria
synaptic knob
vesicle
Ca^{2+} ions
synaptic cleft
receptor site
channel protein opened by receptor protein
Na^+
transmitter molecules
pre-synaptic membrane
post-synaptic membrane
receptor protein
1
2
3

As an action potential arrives at a synaptic knob the following sequence takes place.

- **Channel proteins** in the **pre-synaptic membrane** open to allow Ca^{2+} ions from the synaptic cleft into the synaptic knob.
- **Vesicles** then merge with the pre-synaptic membrane, so that **transmitter molecules** such as **acetylcholine** are **secreted** into the gap.
- The transmitter molecules diffuse across the cleft and bind with specific **sites** in **receptor proteins** in the **post-synaptic membrane**.
- Every receptor protein then opens a **channel protein** so that ions such as Na^+ pass through the post-synaptic membrane into the cell.
- The Na^+ ions **depolarise** the post-synaptic membrane.
- If enough Na^+ ions enter then depolarisation reaches a **threshold level** and an **action potential** is generated in the cell.
- Enzymes in the cleft then remove the transmitter substance from the binding sites, e.g. **acetylcholine esterase** removes **acetylcholine** by hydrolysing it into choline and ethanoic acid.
- Breakdown products of transmitter substances are absorbed into the synaptic knob for re-synthesis of transmitter.

Remember that the generation of an action potential is ALL OR NOTHING. Either enough Na^+ ions pass through the post-synaptic membrane and an action potential is generated OR not enough reach the other side, and there is no effect.

Summation

A single action potential may arrive at a synaptic knob and result in some transmitter molecules being secreted into a cleft. However, there may not be enough to cause an action potential to be generated. If a series of action potentials arrive at the synaptic knob then the build up of transmitter substances may reach the threshold and the neurone will now send an action potential. We say that the neurone has 'fired' as the action potential is produced.

KEY POINT

Once you have understood how an excitatory synapse operates then understanding of the neuromuscular junction should pose no problem!

Look out for questions based on how **drugs** act at synapses. Some **drug molecules** mimic **substances in organisms**. Be ready to compare the molecules of both. Spot the parts of the molecules which are similar. These are the active components!

How do motor neurones control muscle tissue?

The link to muscle tissue is by **motor end plates** which have close proximity to the sarcoplasm of the muscle tissue. The motor end plates have a greater surface area than a synaptic knob, but their action is very similar to the synaptic transmission described on the previous page. Action potentials result in muscle contraction.

Transmitter substances

The synapse described on the previous page used acetylcholine as a transmitter substance. There are others!

- Acetylcholine – the transmitter in the parasympathetic nervous system.
- Noradrenaline – the transmitter in the sympathetic nervous system.
- Dopamine – a transmitter in the brain.
- Serotonin – a transmitter in the brain.

Progress check

1   Explain the importance of summation at a synapse.

2   The diagram shows a synaptic knob.

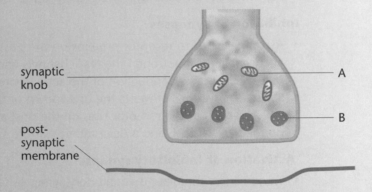

synaptic knob

post-synaptic membrane

A

B

(a) Name A and B

(b) Describe the events which take place after an action potential reaches a synaptic knob and a further action potential is generated as a result.

2 (a) A – mitochondria; B – vesicle

(b) Ca²⁺ ions flow into the synaptic knob; transmitter molecules such as acetylcholine are secreted into the gap; the transmitter molecules bind with sites in receptor proteins in the post-synaptic membrane; this opens channel proteins so that ions such as Na⁺ pass through the post-synaptic membrane into the cell; the post-synaptic membrane is depolarised; if *enough Na⁺ ions enter* a threshold level is reached and an action potential is generated in the cell.

1 A single action potential may arrive at a synaptic knob; there may not be enough transmitter molecules being secreted into a cleft to cause an action potential to be generated; a series of action potentials arrive at the synapse to build up transmitter substances to reach the threshold; the neurone will now send an action potential.

3.3 The effects of drugs on synapses

After studying this section you should be able to:

- *understand the effects of a range of drugs on synapses*
- *understand the mechanisms of amplification and inhibition*

How do drugs affect synaptic transmission?

AQA A — M6
AQA B — M4
EDEXCEL — M4
OCR — M5
WJEC — M5

Different types of drugs can act on synapses in different ways.

Amplification at synapses

- **Amphetamines** stimulate the increase in secretion of noradrenaline so that transmission across the synapse takes place more quickly. Not surprisingly, in the world of illegal drugs it is known as 'Speed'.
- **Caffeine** or **nicotine** amplify synaptic transmission so that less transmitter molecules are needed to stimulate an action potential. (They work by **reducing the threshold** in the post-synaptic membrane.)

Inhibition at synapses

- **Atropine** binds to receptor proteins on the post-synaptic membrane so that acetylcholine cannot open the Na⁺ channels and depolarisation cannot take place. The further conduction of the action potential is stopped.
- **Sarin**, a nerve gas, prevents the breakdown of the transmitter substances, which remain in the receptor sites on the post-synaptic membranes. This has no medical value, being a devastating warfare substance.

Activation of inhibitory synapse

- Tranquillisers stimulate the inhibitory synapses which prevent excitatory synapses from functioning efficiently. Bodily movement slows and perception is less responsive.

A range of drugs have an effect at the synapses.

Some drugs affect the CNS but you will need to remember that a drug is any substance which affects the functioning of the body. The examples in this chapter affect the nervous system but there are many more types.

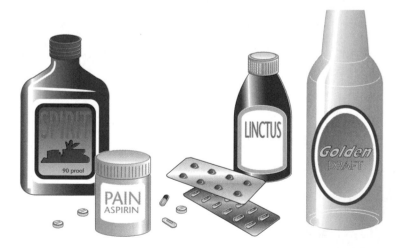

- **Painkillers** stop the conduction of an action potential in the following neurone. Morphine is such a drug so that the feeling of pain is greatly inhibited.
- **Stimulants** like amphetamines speed up synaptic transmission. Increased perception and movements are the result.
- **Depressants** like barbiturates and alcohol slow down conduction across the synapses so that slower reactions and limited perception result.

3.4 The central nervous system

After studying this section you should be able to:

- *outline the structure and functions of the brain and spinal cord*
- *understand the main functions of cerebrum, cerebellum, medulla oblongata and hypothalamus*
- *understand how neurones function together in a reflex arc*
- *describe the structure of skeletal muscle and understand the sliding filament mechanism*
- *outline the features of the autonomic nervous system*
- *describe the structure and function of the eye and ear*

LEARNING SUMMARY

The structure and functions of the CNS

AQA A	M6
AQA B	M4
EDEXCEL	M4
OCR	M5
WJEC	M5
NICCEA	M4

The CNS consists of the brain and spinal cord which work together to aid the coordination of the organism. The human brain has many functions. The spinal cord takes impulses from the brain to **effectors** and in the opposite direction impulses from **receptors** are channelled to the brain.

The brain has a complex 3D structure. The diagram below shows part of the brain structure – major components only.

The CNS is like a motorway with impulses going in both directions!

- **Afferent neurones** take impulses **from** organs **to** the **CNS**.
- **Efferent neurones** take impulses **from** the **CNS** **to** an **organ**.

Learn these carefully. There are no marks for reversal!

The human brain

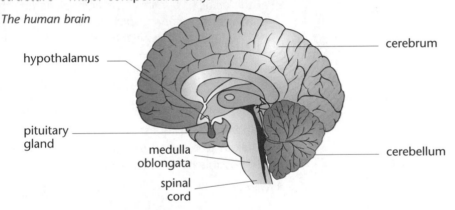

hypothalamus

pituitary gland

medulla oblongata

spinal cord

cerebrum

cerebellum

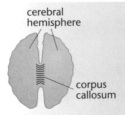

cerebral hemisphere

corpus callosum

There are two cerebral hemispheres: the left and the right. Note that the right hemisphere controls the left side of the body and vice versa.

Functions of parts of the brain

Cerebrum

- **Receives sensory information** from many organs, e.g. impulses are sent from the eyes to the visual cortex at the back of the cerebrum.
- **Controls motor activity** of many organs.
- The front of the cerebrum holds the **memory** and **intelligence** in a network of multi-polar neurones.

Alzheimer's disease

Neurones in the cortex of the cerebrum become progressively less able to produce neurotransmitter substances. Acetylcholine and noradrenaline are usually deficient resulting in major personality changes. The cause is often unknown, but can be genetic.

Cerebellum

- Has a key role in the coordination of **balance** and smooth, controlled muscular movements.
- Initiation of a movement may be by the cerebrum but the **smooth, well-coordinated** execution of the movement is only possible with the help of the cerebellum.

Medulla oblongata

- Its respiratory centre controls the rhythm of breathing with nerve connections to the intercostal muscles and the diaphragm.
- Its cardiovascular centre controls the cardiac cycle via the sympathetic and vagus nerve.
- Connects to the sino-atrial node of the heart.

Hypothalamus

The hypothalamus is the key structure in maintaining a homeostatic balance in the body. It is like a thermostat in a house, switching the heating system on or off as internal conditions change. Similarly it is able to control chemical levels in the blood.

- Has an exceptional blood supply.
- Many receptors are located in the blood vessel walls which supply it.
- These receptors are highly sensitive detectors which monitor:
  - temperature
  - carbon dioxide
  - glucose
  - ionic concentration of plasma.
- Controls the production of thyroxine via the release factor, thyrotrophic hormone.
- Controls ADH secretion by the pituitary gland and is, therefore, responsible for the water content of both blood plasma and urine.

Pituitary gland

- Secretes a range of hormones and release factors and is the major control agent of the endocrine system.
- Responds to neurosecretion by the hypothalamus.
- Together with the hypothalamus is part of a number of negative feedback loops.
- Is the link between the nervous system and the endocrine system.

The above cover some functions of parts of the brain. There are many more!

> The human brain consists of approximately 10^{12} neurones and all are present at birth. It is no wonder that a baby's head is proportionally large at this stage of our life cycle. During the first three months after birth many synaptic connections are made. This is a most important developmental stage. Neurones cannot be replaced once damaged!
>
> **KEY POINT**

The reflex arc

How can we react quickly without even thinking about making a response?

The answer to this question is that the brain is not involved in the response so the time taken to respond to a stimulus is reduced. This rapid, automatic response is made possible by the **reflex arc**.

A reflex arc

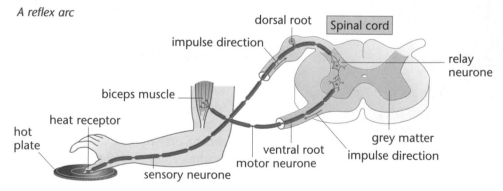

Features of a reflex arc

- The stimulus elicits a response in a **receptor** of a sensory neurone.
- As a result an **action potential** is generated along this sensory neurone.
- The **sensory neurone**, enters the spinal cord via the **dorsal root** and **synapses** onto a relay **neurone**.
- This intermediate neurone synapses onto a **motor neurone** which in turn conducts the impulse to **a muscle** via its **motor end plates**.
- The muscle contracts and the arm instantly **withdraws** from the stimulus before any harm is done.
- The complete list of events takes place so quickly because the impulses do not, initially, go to the brain! The complete pathway to the muscle conducts the impulse so rapidly, before the brain receives any sensory information.

It is other afferent neurones which *finally* take impulses to the brain which enable us to be aware of the arc which has just taken place. This afferent neurone is NOT part of the reflex arc.

> Reflexes have a high survival value because the organism is able to respond so rapidly. Additionally they are always automatic. There are a range of different reflexes, e.g. iris/pupil reflex, knee-jerk reflex and saliva production.

The iris/pupil reflex

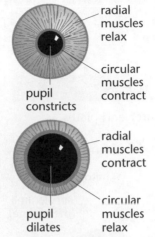

radial muscles relax

circular muscles contract

pupil constricts

radial muscles contract

circular muscles relax

pupil dilates

The diagrams in the margin show the two extremes of pupil size.

- The amount of light entering the eye is detected by **receptors** in the **retina**.
- Reflex pathways lead to the **circular** and **radial** muscles of the **iris**.
- **High intensity** light activates the **circular muscles** of the iris to **contract**, as the radial muscles relax so the **pupil gets smaller**. (The advantage of this is too much light does not enter which would damage the retina.)
- **Low intensity** light activates the **radial muscles** of the iris to **contract**, as the circular muscles relax so the **pupil gets wider**. (The advantage of this is that the eye allows enough light to see.)
- A balance between the two extremes is achieved across a gradation of light conditions.

Autonomic nervous system

This is the part of the nervous system which controls our involuntary activities, e.g. the control of the heart rate. It is divided into two parts, the **sympathetic system** and the **parasympathetic system**. Each system has a major nerve from which smaller nerves branch out into key organs. In some ways the two systems are **antagonistic** to each other but in other ways they have specific functions not counteracted by the other. The table below shows all of the main facts for each system.

This table of features shows some key points for the autonomic system. ALERT! They are difficult to learn because of the lack of logic in the 'pattern' of functions. Take time to revise this properly because many candidates mix up the features of one system with another.

	Autonomic nervous system	
Feature	*Sympathetic*	*Parasympathetic*
Nerve	sympathetic nerve	vagus nerve
Transmitter substance at synapses	noradrenaline	acetylcholine
Heart rate	speeds up	slows down
Iris control	dilates	constricts
Saliva	——————	flow stimulated
Gut movements	slowed down	speeds up
Sweating	sweat production stimulated	——————
Erector pili muscles	contracts erector pili muscles	——————

> Remember that all of the above functions take place without thought. The system is truly involuntary.

Progress check

State **two** functions of each of the following parts of the human brain:

(a) cerebrum (b) cerebellum (c) medulla oblongata.

(a) (i) receives sensory impulses from the eyes to the visual cortex, enabling sight
(ii) controls voluntary motor activity of the leg muscles.
(b) (i) coordinates balance, e.g. enables upright stance in humans
(ii) enables smooth movement, e.g. hitting a golf ball with a club straight down the fairway. (You could swing the club by voluntary control from the cerebrum but smooth coordination is by the cerebellum.)
(c) (i) controls the rhythm of breathing with nerve connections to the intercostal muscles and the diaphragm
(ii) controls the heart rate via the sympathetic and vagus nerve.

3.5 Control of skeletal muscle

After studying this section you should be able to:

● *understand control via motor end plates*
● *understand the mechanism of sarcomere contraction*

LEARNING SUMMARY

How do motor neurones control skeletal muscle?

AQA A	M6
AQA B	M4
EDEXCEL	M4
OCR	M5
WJEC	M5
NICCEA	M4

Motor neurones control the skeletal muscle via motor end plates.

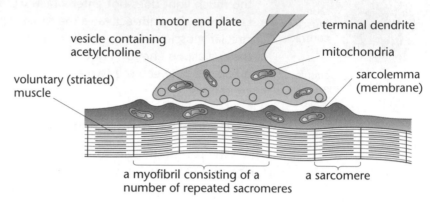

> No contraction would take place without acetylcholine transmitter being released from the motor end plate. When the sarcolemma (membrane) reaches the threshold level, then the action potential is conducted throughout the sarcoplasm. Contraction is initiated!

The skeletal muscles move the bones via their tendon attachments. The muscles work in **antagonistic** pairs, i.e. **opposing** each other. In the arm when the biceps contracts the forearm is lifted. At the same time the triceps relaxes. If the forearm is to be lowered then the triceps contracts and the biceps now relaxes.

Skeletal muscle is also known as striated or striped muscle. The structure of a single muscle unit, the sarcomere, shows the striped nature of the muscle.

The sarcomere

A sarcomere showing bands   *A sarcomere showing filaments*

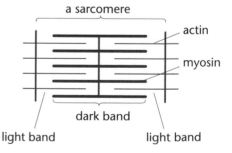

- The sarcomere consists of different filaments, thin ones (**actin**) and thick ones (**myosin**).
- These filaments form bands of different shades:
  - light band – just actin filaments
  - dark band – just myosin filaments or myosin plus actin.
- During contraction the filaments slide together to form a shorter sarcomere.
- As this pattern of contraction is repeated through 1000s of sarcomeres the whole muscle contracts.
- Actin and myosin filaments slide together because of the formation of cross bridges which alternatively build and break during contraction.
- Cross bridge formation is known as the 'ratchet mechanism'.

How does the 'ratchet mechanism' work?

To answer this question the properties of actin and myosin need to be considered. The diagram below represents an actin filament next to a myosin filament. Many 'bulbous heads' are located along the myosin filaments (just one is shown!). Each points towards an actin filament.

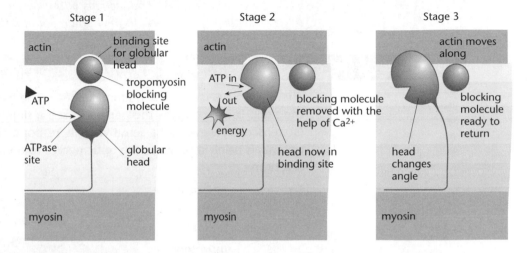

The sequence of the ratchet mechanism

- Once an action potential is generated in the muscle tissue then **Ca^{2+} ions** are released from the **reticulum**, a structure in the **sarcoplasm**.
- Part of the **globular head** of the myosin has an **ATPase** (enzyme!) site.
- Ca^{2+} ions activate the myosin head so that this ATPase site hydrolyses an ATP molecule, **releasing energy**.
- Ca^{2+} ions also bind to **troponin** in the actin filaments, this in turn removes **blocking molecules (tropomyosin)** from the actin filament.
- This exposes the **binding sites** on the **actin** filaments.
- The globular heads of the myosin then bind to the newly exposed sites forming **actin-myosin cross bridges**.
- At the stage of energy release the myosin **heads change angle**.
- This change of angle moves the actin filaments towards the centre of each sarcomere and is termed the **power stroke**.
- More ATP binds to the myosin head, effectively causing the cross bridge to become straight, the tropomyosin molecules once again block the actin binding sites.
- The myosin is now 'cocked' and ready to repeat the above process.
- Repeated cross bridge formation and breakage results in a rowing action shortening the sarcomere as the filaments slide past each other.

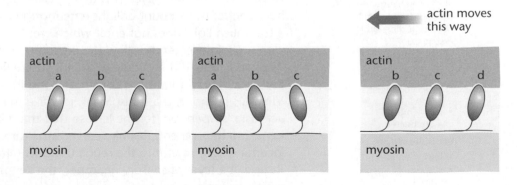

3.6 The human eye and ear

After studying this section you should be able to :

- *describe structures and their functions in the human eye*
- *describe the structures and functions of the cells in the retina*
- *describe the key structures in the human ear and their functions*

LEARNING SUMMARY

Structure of the eye

AQA A	M6
AQA B	M4
EDEXCEL	M4
OCR	M5
NICCEA	M4

The human eye is a typical mammalian eye. The two eyes are located in sockets in the skull. Each eye helps the person to see a slightly different image giving a 3D (stereoscopic) view. Light sensitive cells respond to the light resulting in action potentials being sent along the optic nerves to the visual cortex at the back of the cerebrum.

The sensory nerve connections of the eyes

right eye

left eye

optic chiasma where nerves cross over

Note that each visual field is slightly different. The routes of the nerves crossing over help to create 'depth' in the image.

Section through a human eye

ciliary body

suspensory ligaments

cornea

pupil

aqueous humour

iris

conjunctiva

sclera

vitreous humour

optic nerve

blind spot

fovea

retina

choroid

Key structures and their functions

The key parts in the focusing of light are the cornea and lens which changes shape when focusing near or far objects.

Accommodation is the ability to change the focal point of the lens to focus objects at different distances.

Near object focus:
- lens – round
- suspensory ligaments – loose
- ciliary muscle – contracted.

Far object focus:
- lens – long and thin
- suspensory ligament – tight
- ciliary muscle – relaxed

- **Conjunctiva**, is a thin protective covering of epithelial cells.
- **Cornea**, is the transparent, curved front of the eye which helps to converge the light rays which enter the eye.
- **Sclera**, is an opaque, fibrous, protective outer structure.
- **Choroid**, has a network of blood vessels to supply nutrients to cells and remove waste products. It is pigmented to prevent internal reflection.
- **Iris**, is a pigmented muscular structure consisting of an inner ring of **circular muscle** and an outer layer of **radial muscle** (see page 55). Its function is to help control the amount of light entering the eye so that:
  – too much light does not enter which would damage the retina
  – enough light enters to allow the person to see.
- **Pupil**, is a hole in the middle of the iris where light is allowed to continue its passage. In bright light it is constricted and in dim light it is dilated.
- **Vitreous humour**, is a transparent, jelly-like mass located behind the lens. It acts as a 'suspension' for the lens so that the delicate lens is not damaged.
- **Lens**, is a transparent, flexible, curved structure. Its function is to focus incoming light rays onto the retina using its refractive properties.
- **Retina**, is a layer of sensory neurones, the key structures being photoreceptors which respond to light.
- **Blind spot**, is where the bundle of sensory fibres form the optic nerve.

How do the cells of the retina respond to light?

There are two types of cell which are **photosensitive**, the rod cells and the cone cells. They each have different properties.

Rod cells

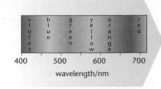

- Are very sensitive to the **intensity of light**, but are not sensitive to colour.
- Can respond to even dim light.
- Respond by the following reaction:

$$\text{rhodopsin} \xrightarrow{\text{light}} \text{opsin} + \text{retinal}$$

- Opsin opens **ion channels** in the cell surface membrane which can result in the generation of an action potential.
- Rhodopsin can be re-generated during an absence of light.

Cone cells

- Require **high light intensities** to be responsive.
- Respond by the following reaction:

$$\text{iodopsin} \xrightarrow{\text{light of specific wavelength}} \text{photopsin} + \text{retinal}$$

The colour vision mechanism appears to require three types of cone, RED, GREEN, and BLUE and gives the **trichromatic** theory.

- Exist in three different types, red, green and blue, each having a different form of iodopsin:
    – **red** cones are stimulated by wavelengths of red light
    – **green** cones are stimulated by wavelengths of green light
    – **blue** cones are stimulated by wavelengths of blue light
    – all three cones when stimulated give white light
    – none of the three types when stimulated gives black.
- Opsin again opens **ion channels** in the membranes which can lead to the generation of an action potential.

Structure of the retina

Remember that red light reaching a cone sensitive to only blue light would not stimulate the generation of an action potential! Cones are only sensitive to light of a specific range of wave-lengths.

bipolar cell — synapse — fibre of optic nerve

direction of light

synapse

nucleus

mitochondria

outer segment (with rhodopsin)

Note the direction of light as it reaches the upper retinal surface.

pigment cell

Rod cells

bipolar cell — synapse — fibre of optic nerve

direction of light

outer segment with iodopsin (photosensitive pigment)

A cone cell

Visual acuity

This is a measure of the detail we can see. The **cones** are responsible for high visual acuity (**high resolution!**). **Large numbers** are packed **very close** to each other in the fovea. **ONE** cone cell synapses onto **ONE** bipolar cell which in turn synapses onto **ONE** ganglion cell as the information is relayed to the visual cortex. Spatially, much more clarity is perceived than for the rods. The image can be likened to a television picture with high numbers of pixels. High definition!

Compare this with the rods. The rod cells are not packed close together so that visual acuity is low.

Did you know that the fovea consists of almost totally cones?

What do we see with in dim conditions?

You have guessed it, rods. Their ability to detect dim light is useful but there is no colour!

Convergence

Many rods can synapse onto one bipolar cell. A ray of light reaching one rod may not be enough to stimulate an action potential along a nerve pathway. **Several** rods link to **one** bipolar cell so that enough transmitter molecules at a synapse reach the threshold level. This depolarisation results in an action potential in the bipolar cell. This is **summation**, as a result of rod cell teamwork!

Structure of an ear

OCR ▷ M5
WJEC ▷ M5

The human ear responds to sound frequencies within the range 40–20 000 Hz. Outside of this auditory range the sensory neurones do not respond.

Section through a human ear

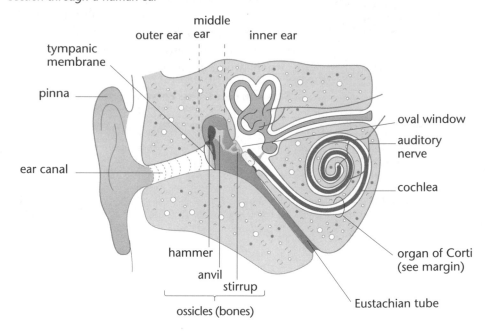

The Eustachian tube allows the pressure at either side of the tympanic membrane to be equalised, i.e. atmospheric pressure = pressure in the inner ear.

Key structures and their functions

- **Sound waves** are directed towards the tympanic membrane through the ear canal.
- They reach the **tympanic membrane** which vibrates as a result.
- The **ossicle bones** then **pass on** and **amplify** these vibrations.
- The membrane of the **oval window** then vibrates and passes on these vibrations to the **perilymph** of the **cochlea**.
- Movements of the fluid perilymph pass through the cochlea and move **sensory hair cells** on the **organ of Corti**.
- Physical hair movements result in the generation of action potentials which pass along the **auditory nerve** (sensory hairs are transducers).

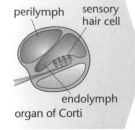

perilymph   sensory hair cell

endolymph
organ of Corti

3.7 Plant sensitivity

After studying this section you should be able to:

- *understand the range of tropisms which affect plant growth*
- *understand how auxins, gibberellins and cytokinins control plant growth*
- *understand how phytochromes control the onset of flowering in plants*

LEARNING SUMMARY

Plant growth regulators

EDEXCEL	M5
OCR	M5
WJEC	M5
NICCEA	M4

External stimuli such as light can affect the direction of plant growth. A **tropism** is a **growth response** to an external stimulus. It is important that a plant grows in a direction which will enable it to obtain maximum supplies. Plant regulators are substances produced in minute quantities and tend to interact in their effects.

> Growth response to light is **phototropism**
> Growth response to gravity is **geotropism**
> Growth response to water **hydrotropism**
> Growth response to contact is **thigmotropism**
> Tropisms can be positive (towards) or negative (away from).
>
> **KEY POINT**

Phototropism

This response is dependent upon the stimulus, light affecting the growth regulator, **auxin** (indoleacetic acid).

Auxin and growing shoots

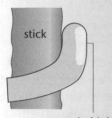

stick

auxin high
concentration here
so cells elongate

Thigmotropism helps a climbing plant like the runner bean to grow in a twisting pattern around a stick. Auxin is redistributed away from the contact point so the outer cells elongate giving a stronger outer growth.

Auxin is produced by cells undergoing mitosis, e.g. growing tips. If a plant shoot is illuminated from one side then the auxin is redistributed to the side furthest from the light. This side grows more strongly, owing to the elongation of the cells, resulting in a bend towards the light. The plant benefits from increased light for photosynthesis. Up to a certain concentration the degree of bending is proportional to auxin concentration.

Tropisms in response to light light from different directions

Tropism in response to auxin

The diagrams show tropic responses to light and auxin.

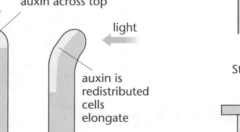

☐ auxin

light

equal amount of auxin across top

light

auxin is redistributed cells elongate

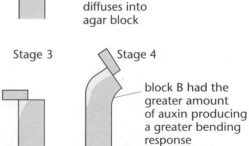

Stage 1
tip cut off

Stage 2
tip placed on agar block

auxin diffuses into agar block

Stage 3

Stage 4
block B had the greater amount of auxin producing a greater bending response

Auxin research

Many investigations of auxins have taken place using the growing tips (coleoptiles) of grasses. Where a growing tip is removed and placed on an agar block, auxin will diffuse into the agar. Returning the block to the mitotic area stimulates increased cell elongation to the cells receiving a greater supply of auxin.

Is the concentration of auxin important?

It is important to consider the implications of the concentration of auxin in a tissue. The graph below shows that at different concentrations auxin affects the shoot and the root in different ways.

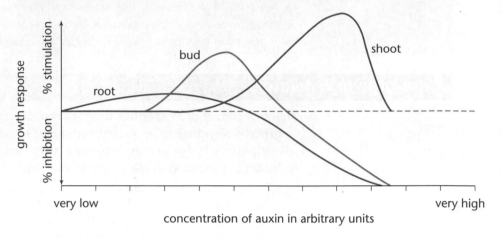

Analyse this graph carefully. It shows how the same substance can both stimulate or inhibit, depending on concentration.

The graph shows:

- auxin has no effect on a shoot at very low concentration
- at these very low concentrations root cell elongation is stimulated
- at higher concentrations the elongation of shoot cells is stimulated
- at these higher concentrations auxin inhibits the elongation of root cells.

Auxin and root growth

The graph shows that auxin affects root cells in a different way at different concentrations. At the root tip auxin accumulates at a lower point because of gravity. This inhibits the lower cells from elongating. However, the higher cells at the tip have a low concentration of auxin and do elongate. The net effect is for the stronger upper cell growth to bend the root downwards. The plant therefore has more chance of obtaining more water and mineral ions.

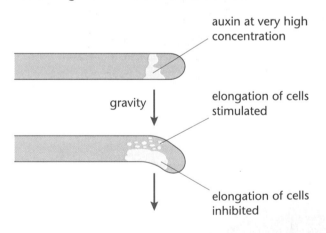

Plant growth regulators

In your examination look out for data which will be supplied, e.g. the growth regulator gibberellin may be linked to falling starch levels in a seed endosperm and increase in maltose. Gibberellin has stimulated the enzymic activity.

Hormone	Some key functions
auxin	increased cell elongation, suppression of lateral bud development
gibberellin	cell elongation, ends dormancy in buds, promotes germination of seeds by activating hydrolytic enzymes such as amylase (food stores are mobilised!)
cytokinin	increased cell division, increased cell enlargement in leaves
ethene	promotes ripening of food

Phytochrome and the onset of flowering in plants

Phytochrome:

- is a regulatory substance
- is photosensitive
- has a number of different roles
- controls the onset of flowering in plants
- exists in two different forms, **phytochrome red** (P_R) and **phytochrome far red** (P_{FR}).

The two forms are inter-convertible as shown below.

The terms P_R and P_{FR} refer to the peak wavelengths of light, absorbed by each substance:

P_R absorbs a peak of 665 nm

P_{FR} absorbs a peak of 725 nm.

very fast in red light
fast in sunlight

$$P_R \quad \underset{\longleftarrow}{\overset{\longrightarrow}{}} \quad P_{FR}$$

slow conversion in dark
fast in far red light

The above inter-conversion of the phytochromes is part of the mechanism that controls the onset of flowering. This is known as **photoperiodism**.

Different species of plants respond to different day lengths during the year.

Specific day length triggers the development of the flower buds.

There are three categories of plant:

- long day plants, e.g. petunias (need P_{FR} to flower)
- short day plants, e.g. chrysanthemums (need P_R to flower)
- day neutral plants, e.g. tomatoes.

The day-length and night-length bars below show the proportion of light and dark and the effect on the flowering.

long day plants in summer

- $P_R \longrightarrow P_{FR}$ (fast conversion in light)
- flowering promoted (long day plants need P_{FR})

short day plants in summer

- $P_R \longrightarrow P_{FR}$
- flowering NOT promoted (short day plants need P_R)

In your examinations look out for data about day length. Particularly look for the short day data where a flash of light occurs during a dark period. This is enough to make P_{FR} which will stimulate long day plants to flower. This is due to the rapid $P_R \rightarrow P_{FR}$ process.

long day plants in winter

- $P_{FR} \longrightarrow P_R$ (slow process but the night is long enough)
- flowering not promoted (long day plants need P_{FR})

short day plants in summer

- $P_{FR} \longrightarrow P_R$ (slow process but long enough at night to make P_R)
- flowering promoted (short day plants need P_R)

Key

darkness

sunlight

Sample question and model answer

The diagram below shows a section through a mammalian eye.

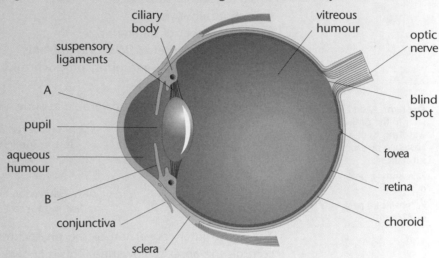

Take care when answering questions like this! Initially it could be a GCSE question but the level of detail required is much greater. Use the key technical terms, i.e. the fact that the cornea is able to begin the focusing of light rays by convergence, possible because of its refractive properties.

Similarly it is important to give detail of the radial and circular muscles of the iris to regulate light entry.

(a) Describe the function of parts A and B. [2]

A (cornea) refracts the light to begin convergence of light rays entering the eye.

B (iris) the radial and circular muscles of the iris change the size of the pupil to regulate light entry.

(b) Complete the table below to show the function of each part in the table.

	lens	suspensory ligament	ciliary body	
Focusing a near object	round	loose	contracted	[1]
Focusing a far object	longer and thinner	tight	relaxed	[1]

(c) The diagram shows a cone from the retina.

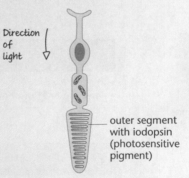

Remember the position of a cone in the retina. This will help you decide that the light comes down towards this cone. Learn this carefully.

(i) Place an arrow on the diagram to show the direction in which light reaches the cone. [1]

Always look at how many marks per question part. You then know how many points to make.

(ii) What is the function of the iodopsin in the outer segment? [4]

when stimulated by light of the correct wavelength – breaks down to release opsin; opsin opens ion channels in the membranes; this can lead to the generation of an action potential; in a bipolar cell.

(iii) How do cones contribute to high visual acuity? [3]

cones are tightly packed giving a high surface area; each cone synapses onto a single bipolar neurone; so the greater detail gives higher resolution.

Practice examination questions

1 The growing tips (coleoptiles) were removed from oat stems. Agar blocks containing different concentrations of synthetic auxin (IAA) replaced the tips on the oat stems. The plants were allowed to grow for a period then the angle of curvature of the stems was measured. The results are shown in the graph below.

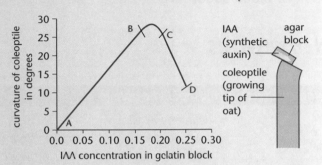

(a) What is the relationship between IAA concentration and curvature of the stem between points:

    (i) A and B   [1]

    (ii) C and D?   [1]

(b) Explain how IAA causes a curvature in the oat stems.   [2]

(c) Explain the effect a much higher concentration of IAA would have on the curvature of oat stems.   [2]

                                                          [Total: 6]

2 The diagram below shows a single sarcomere just before contraction.

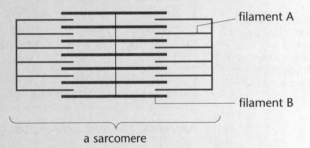

(a) Name filaments A and B.   [2]

(b) What stimulus causes the immediate contraction of a sarcomere?   [1]

(c) What happens to each type of filament during contraction?   [2]

                                                          [Total: 5]

3 The diagram below shows the profile of an action potential.

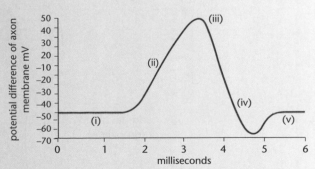

Explain what happens in the axon at each stage shown on the diagram.   [10]

                                                           [Total: 10]

Chapter 4

Homeostasis

The following topics are covered in this chapter:

- Hormones
- Temperature control in a mammal
- Regulation of blood sugar level
- The kidneys
- Adaptations to desert ecosystems

4.1 Hormones

After studying this section you should be able to:

- *define homeostasis*
- *describe the route of hormones from source to target organ*
- *understand how hormones contribute to homeostasis*
- *recall the roles of a range of hormones*

The endocrine system

AQA A	M6
AQA B	M4
EDEXCEL	M4
OCR	M4
WJEC	M5
NICCEA	M4

The endocrine system secretes a number of chemicals known as **hormones**. Each hormone is a substance produced by an **endocrine gland**, e.g. adrenal glands produce the hormone adrenalin. Every hormone is **transported in the blood** and has a **target organ**. Once the target organ is reached the hormone **triggers a response** in the organ. Many hormones do this by **activating enzymes**. Others **activate genes**, e.g. steroids.

The great advantage of homeostasis is that the conditions in the environment fluctuate but conditions in the organism remain stable.

> **KEY POINT**
>
> The endocrine and nervous systems both contribute to **coordination** in animals. They help to regulate internal processes. **Homeostasis** is the maintenance of a **constant internal environment**. Nerves and hormones have key roles in the maintenance of this **steady internal state**. Levels of pH, blood glucose, oxygen, carbon dioxide and temperature all need to be controlled.

Parts of the human endocrine system (both male and female organs shown!)

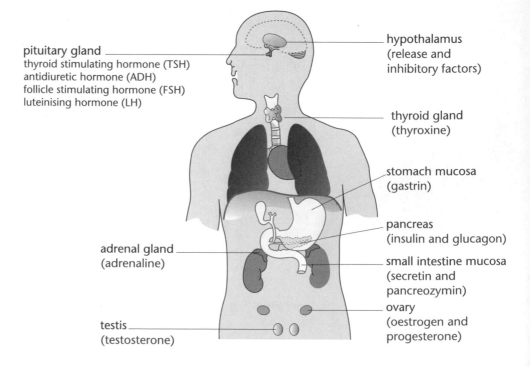

pituitary gland
thyroid stimulating hormone (TSH)
antidiuretic hormone (ADH)
follicle stimulating hormone (FSH)
luteinising hormone (LH)

hypothalamus
(release and
inhibitory factors)

thyroid gland
(thyroxine)

stomach mucosa
(gastrin)

pancreas
(insulin and glucagon)

adrenal gland
(adrenaline)

small intestine mucosa
(secretin and
pancreozymin)

ovary
(oestrogen and
progesterone)

testis
(testosterone)

Some important mammalian hormones

The table shows the sources and some functions of a range of mammalian hormones. The pituitary gland is the key control gland. It affects many areas of the body and even stimulates, by the production of **tropic hormones**, other hormones, e.g. follicle stimulating hormone.

Hormone	Source	Effect
TSH	pituitary	stimulates the thyroid to secrete thyroxine
thyroxine	thyroid	increases metabolic rate
insulin	pancreas	reduces blood sugar
glucagon	pancreas	stimulates conversion of glycogen to glucose in the liver
ADH	pituitary	increased water reabsorption by kidney
gastrin	stomach mucosa	stimulates HCl production in stomach
secretin	intestinal mucosa	stimulates the pancreas to secrete fluid + alkali
pancreozymin	intestinal mucosa	stimulates the pancreas to secrete enzymes
FSH	pituitary	stimulates primary (Graafian) follicle to develop or testis to make sperms
LH	pituitary	stimulates ovulation or testis to make testosterone
oestrogen	ovary	stimulates development of endometrium stimulates secondary sexual characteristics
progesterone	ovary	maintains endometrium

How does a hormone trigger a cell in a target organ?

Hormones are much slower in eliciting a response than the nervous system. Rather than having an effect in milliseconds like nerves, hormones take longer. However, effects in response to hormones are often **long lasting**.

The diagram below shows one mechanism by which hormones activate target cells.

Did you know?
Each enzyme shown is constantly re-used as an active site is left free.

Look carefully at this mechanism! Just **ONE hormone molecule** arriving at the cell releases an enzyme which can be used **many** times. In turn, another enzyme is produced which can be used **many** times. One hormone molecule leads to **amplification**. This is a cascade effect!

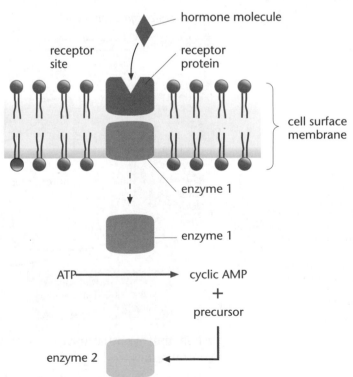

4.2 Temperature control in a mammal

LEARNING SUMMARY

After studying this section you should be able to:

- *outline the processes which contribute to temperature regulation in a mammal*
- *understand how nervous and endocrine systems work together to regulate body temperature*
- *understand how internal processes are regulated by negative feedback*

What are the advantages of controlling body temperature?

AQA A	M6
AQA B	M4
EDEXCEL	M4
OCR	M5
NICCEA	M4

It is advantageous to maintain a constant body temperature so that the enzymes which drive the processes of life can function at an optimum level.

- **Endothermic** (warm blooded) animals can maintain their core temperature at an optimal level. This allows internal processes to be consistent. The level of activity of an endotherm is likely to fluctuate less than an ectotherm.

- **Ectothermic** (cold blooded) animals have a body temperature which fluctuates with the environmental temperature. As a result there are times when an animal may be vulnerable due to the enzyme driven reactions being slow. You could approach a crocodile (ectotherm) in cold conditions. Its speed of attack would be slow. If approached in warm conditions the attack would be rapid.

Once the blood temperature decreases, the heat gain centre of the hypothalamus is stimulated. This leads to a rise in blood temperature which, in turn, results in the heat loss centre being stimulated. This is negative feedback! The combination of the two, in both directions, contributes to homeostasis.

The **hypothalamus** has **many functions**! It controls thirst, hunger, sleep and it stimulates the production of many hormones other than those required for temperature regulation.

How is temperature controlled in a mammal?

The key structure in homeostatic control of all body processes is the **hypothalamus**. The regulation of temperature involves thermoreceptors in the skin, body core and blood vessels supplying the brain, which link to the hypothalamus.

The diagram below shows how the peripheral nerves, hypothalamus and pituitary gland integrate nervous and endocrine glands to regulate temperature.

Temperature regulation model

If there is an increase in core temperature then the hypothalamus stimulates greater heat loss by:

- vasodilation (dilation of the skin arterioles)
- erector-pili muscles are relaxed and hairs lie flat
- more sweating
- behavioural response in humans to change to thinner clothing.

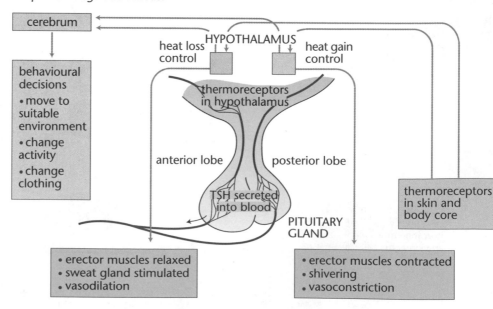

When the hypothalamus receives sensory information **heat loss** or **heat gain** control results.

A **fall** in temperature results in the following control responses.

A capillary bed

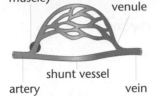

arteriole
(a sphincter
muscle)

venule

shunt vessel

artery

vein

Vasoconstriction

- Arteriole control is initiated by the hypothalamus which results in efferent neurones stimulating constriction of the arteriole sphincters of skin capillary beds.
- This deviates blood to the core of the body, so less heat energy is lost from the skin.

Contraction of the erector-pili muscles

- Erector-pili muscle contraction is initiated in the hypothalamus being controlled via efferent neurones.
- Hairs on skin stand on end and trap an insulating layer of air, so less heat energy is lost from the skin.

Sweat reduction

- The sweat glands control is also initiated in hypothalamus, and is controlled via efferent neurones.
- Less heat energy is lost from the skin by evaporation of sweat.

Shivering

- Increased muscular contraction is accompanied by heat energy release.

Behavioural response

- This could be to switch on the heating, put on warmer clothes, etc.
- A link from the hypothalamus to the cerebrum elicits this voluntary response.

Increased metabolic rate

- The hypothalamus produces a release factor substance.
- This stimulates the anterior part of the pituitary gland to secrete TSH (thyroid stimulating hormone).
- TSH reaches the thyroid via the blood.
- Thyroid gland is stimulated to secrete thyroxine.
- Thyroxine increases respiration in the tissues increasing the body temperature.

Once a higher thyroxine level is detected in the blood the release factor in hypothalamus is inhibited so TSH release by the pituitary gland is prevented. This is **negative feedback**.

An **increase** in body temperature results in almost the **opposite** of each response described for a fall of temperature.

Vasodilation

- Arterioles of capillary beds dilate allowing more blood to skin capillary beds.

Relaxation of erector-pili muscles

- Hairs lie flat, no insulating layer of air trapped.
- More heat loss of skin.

Sweat increase

- More sweat excreted so more heat energy from body needed to evaporate the sweat, so we cool down.

Behavioural response

- This could be to move into the shade or consume a cold drink.

Note:

(a) the outline for heat loss methods does not show the nerve connections. Efferent neurones are again coordinated via the hypothalamus!

(b) heat is lost from the skin via a combination of **conduction**, **convection** and **radiation**.

Progress check

Hormone X stimulates the production of a substance in a cell of a target organ. The following statements outline events which result in the production of the substance but are in the wrong order. Write the correct order of letters.

A Hormone X is transported in the blood.

B Hormone X binds with a receptor protein in the cell surface membrane.

C The enzyme catalyses a reaction, forming a product.

D Hormone X secreted by gland.

E This releases an enzyme from the cell surface membrane.

D, A, B, E, C.

4.3 Regulation of blood sugar level

After studying this section you should be able to:

- *understand the control of blood glucose levels in a person*
- *describe the sites of insulin and glucagon secretion*
- *explain the functions of insulin and glucagon*

Why is it necessary to control the amount of glucose in the blood?

AQA A	M6
AQA B	M4
EDEXCEL	M4
OCR	M5
NICCEA	M4

Glucose molecules are needed to supply energy for every living cell. The level in the blood must be high enough to meet this need (90 mg per 100 cm^3 blood). This level needs to be maintained at a constant level, even though a person may or may not have eaten. High levels of glucose in the blood would cause great problems. Hypertonic blood plasma would result in water leaving the tissues by osmosis. Dehydration of organs would result in a number of symptoms.

Blood glucose regulation

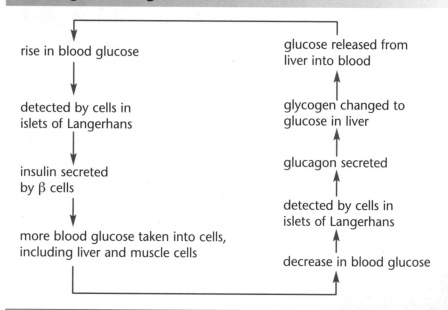

rise in blood glucose → detected by cells in islets of Langerhans → insulin secreted by β cells → more blood glucose taken into cells, including liver and muscle cells

decrease in blood glucose → detected by cells in islets of Langerhans → glucagon secreted → glycogen changed to glucose in liver → glucose released from liver into blood

Negative feedback

Blood glucose regulation is an example of negative feedback. Any change in glucose level initiates changes which will result in the return of the original level, **balance is achieved**.

KEY POINT

Insulin

- Is secreted into the blood due to stimulation of pancreatic cells by a **high concentration** of glucose in the blood.
- Is produced by the β **cells** of the **islets of Langerhans** in the pancreas.
- Binds to receptor proteins in cell surface membranes activating carrier proteins to **allow glucose entry** to cells.
- Allows **excess glucose** molecules into the liver and muscles where they are converted into **glycogen** (a storage product), and some fat.

Never state that insulin changes glucose to glycogen. It allows glucose into the liver where glycogen synthase catalyses the conversion!

Glucagon

- Is secreted into the blood due to stimulation of pancreatic cells by a **low concentration** of glucose in the blood.
- Is produced by the α **cells** of the **islets of Langerhans** in the pancreas.
- Stimulates the conversion of glycogen to **glucose**.

Diabetes

There are two types of this condition.

Type 1

- The pancreas fails to produce enough insulin.
- After a meal when blood glucose level increases dramatically, the level remains high.
- High blood glucose causes hyperglycaemia.
- Kidneys, even though they are healthy, cannot reabsorb the glucose, resulting in glucose being in the urine.
- Symptoms include dehydration, loss of weight and lethargy.

What is the answer?
- Insulin injections and carbohydrate controlled diet.

Type 2

- This form of diabetes usually occurs in later life.
- Insulin is still produced but the receptor proteins on the cell surface membranes may not work correctly.
- Glucose uptake by the cells is erratic.
- Symptoms are similar to those for type 1 but are mild in comparison.

What is the answer?
- Dietary control including low carbohydrate intake.

More liver functions

The role of the liver in its production of bile, as well as the storage and break down of glycogen has been highlighted. The liver does so much more!

Transamination

This is the way an R group of a keto acid is transferred to an amino acid. It replaces the existing R group with another, a new amino acid has been formed.

Children need 10 essential amino acids (adults need 8). From these they can make different ones by transamination in the liver!

CH_3 amino acid + C_2H_5 keto acid → C_2H_5 amino acid + CH_3 keto acid

note the changes in the 'R' group of each acid.

Deamination

This process is necessary to lower the level of **excess amino acids**. They are produced when proteins are digested. Nitrogenous materials have a high degree of toxicity, so the level in the blood must be limited.

The liver has many functions including:
- detoxification of poisonous substances
- heat production
- formation of red cells.

$$2NH_2\text{--}\underset{\underset{H}{|}}{\overset{\overset{R}{|}}{C}}\text{--}COOH + O_2 \rightarrow 2\underset{\underset{O}{||}}{\overset{\overset{R}{|}}{C}}\text{--}COOH + 2NH_3$$

amino acid   oxygen   keto acid   ammonia

Ornithine cycle

Both deamination and the ornithine cycle are needed to process excess amino acids. Remember that urea is a less toxic substance. The **liver makes** it but the **kidneys** help to **excrete** it.

Ammonia is immediately taken up by ornithine to help make a less toxic substance, urea.

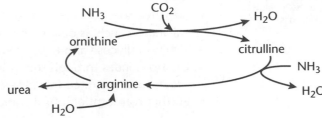

4.4 The kidneys

After studying this section you should be able to:

- describe the structure and functions of a nephron
- understand the processes of ultrafiltration and reabsorption
- understand the countercurrent multiplier

LEARNING SUMMARY

Kidney structure and function

AQA A	M6
AQA B	M4
EDEXCEL	M4
OCR	M5
WJEC	M5
NICCEA	M4

Each kidney has three major regions: the **cortex**, **medulla** and **pelvis**. The renal artery takes blood into a kidney where it is filtered to remove potentially toxic material. Useful substances leave the blood as well as toxic ones but are reabsorbed back into the blood. Toxic substances such as urea leave the kidneys and enter the bladder, via the ureters.

The diagram shows one nephron of the many thousands in each kidney.

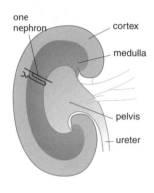

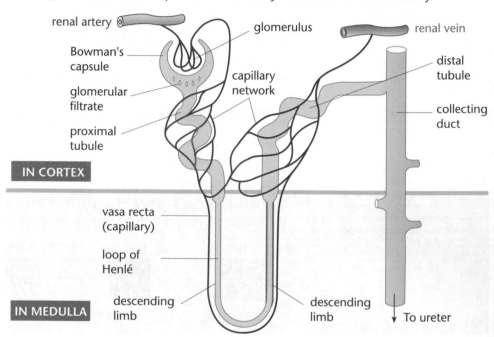

How does a nephron function?

- Blood arrives at the **glomerulus** from the renal artery.
- The **blood pressure is very high** as a result of:
  - contraction of the left ventricle of the heart
  - contraction of aorta and renal artery
  - the arteriole leading to glomerular capillaries is wider than venule leaving them
  - high resistance of the interface between glomerular capillaries and inner wall of the renal (Bowman's) capsule.
- **Glomerular filtrate** is forced into the nephron, this is known as **ultrafiltration**.
- Glomerular filtrate includes **urea, glucose, water, amino acids** and **mineral ions**.
- Selective reabsorption takes place in the **proximal tubule** resulting in substances such as glucose being returned to the blood.
- 100% of glucose and 80% of water are reabsorbed at the proximal tubule.
- Urea continues through the tubule to the collecting duct and finally down a ureter to be excreted from the bladder.
- Further reabsorption of substances can take place at the distal tubule.

The selective property of the renal membrane.

What do not leave the blood due to being too large?

Most proteins, red and white blood cells.

Ultrafiltration

Also known as pressure filtration it relies on the properties of the capillaries and the inner wall of the renal (Bowman's) capsule.

In your examination you may be requested to label a diagram. Test yourself!

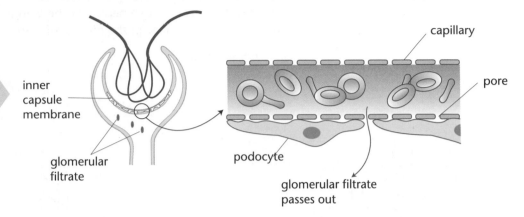

- Capillaries lie very close to the **inner capsular membrane** (see above).
- The capillaries have many tiny pores.
- The capsular membrane consists of **podocytes**.
- **Podocytes** are shaped so that many tiny gaps exist between the capsular membrane and capillaries.
- Together these form a high pressure sieve.
- Only molecules which are small enough can pass through.

Learn explanations in bullet points. Bullet points in this book may resemble the examiner's mark scheme.

Reabsorption

Capillaries from the glomerulus extend to a network across both proximal and distal tubules. The close contact between capillary and tubule is important.

Section through proximal tubule

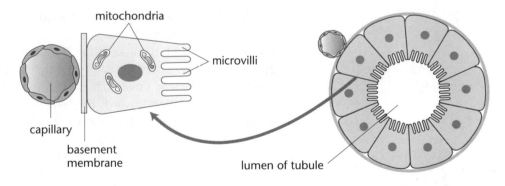

Remember what reabsorption is! The return of substances into the blood which have just left.

- Substances such as **glucose**, **urea**, and **water** travel along the tubule.
- Each tubule is one cell thick, consisting of epithelial cells with **microvilli** on the outer membrane.
- Microvilli give a **high surface area** to allow the efficient transport of substances to cross to the capillaries.
- **Carrier proteins** on the microvilli, aided by mitochondria, **actively reabsorb** glucose from the filtrate into the tubules.
- Glucose molecules are then actively transported into the fluids surrounding the capillaries.
- Glucose molecules finally enter the capillaries and so have re-entered the blood.
- By the end of the proximal tubule **all glucose** has been **returned to the blood**.

The distal tubule is also in close contact with the capillary network. Even more reabsorption can take place here.

How do the kidneys conserve water?

Water molecules which pass into the tubule and reach the kidney pelvis continue down a ureter and are lost in urine. Such water loss is carefully controlled, some is always reabsorbed. This control involves both the nervous system, the endocrine system and structures along a nephron. The diagram below outlines the role of the **countercurrent multiplier** in the control of water content in the body.

Countercurrent multiplier

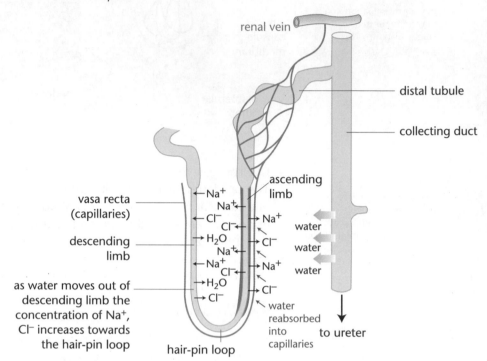

The vasa recta capillaries follow the path of the loop of Henlé to:
(a) supply oxygen to the cells so that active transport of Na^+, Cl^- can take place efficiently (the process needs energy!)
(b) remove CO_2
(c) reabsorb water.

as water moves out of descending limb the concentration of Na^+, Cl^- increases towards the hair-pin loop

The role of the loop of Henlé

- Na^+ and Cl^- ions are actively transported into the medulla from the ascending limb of the loop of Henlé.
- The ascending limb is thicker than the descending one, and impermeable to the outward movement of water so only the ions leave.
- The Na^+ and Cl^- ions slowly diffuse into the descending limb resulting in their greater concentration towards the base of the loop.
- A high concentration of Na^+ and Cl^- ions in the medulla causes water to leave the collecting duct by osmosis.
- Additionally water leaves the descending limb by osmosis due to the ions in the medulla.
- Water molecules pass into the capillary network and have been successfully reabsorbed.

The role of the distal tubule

The distal tubule is also a site of more reabsorption. Even more substances are returned to the blood here.

The structure of the distal tubule is similar to the proximal tubule, however its specific roles are:

- maintenance of a constant blood plasma pH at around 7.4
- if blood plasma falls **below** a pH of 7.4 then ionic movements take place

    (H^+ ions)   plasma $\rightarrow$ filtrate
    (HCO_3^- ions)   filtrate $\rightarrow$ plasma

- if blood plasma **rises** above a pH of 7.4 then more ion movements take place

    (OH^- ions)   plasma $\rightarrow$ filtrate
    (HCO_3^- ions)   plasma $\rightarrow$ filtrate

The control of water balance

It is necessary to control the amount of water in the blood. The kidneys can help to achieve this with their ability to intercept water before it can reach the ureters. There are, however, problems to overcome.

> In hot conditions we lose a lot of water by sweating, too much loss would lead to dehydration problems.
>
> In cold conditions much less water is lost by sweating, giving a potential problem of too much water being retained in the blood.
>
> A balance must be achieved!

KEY POINT

Here the consequences of the two extremes of hot and cold are explained. Do remember that there are a range of conditions **between** these extremes. ADH level changes in response to osmoreceptor sensory input to the hypothalamus.

Hormonal control of the kidneys – the role of ADH

Control is achieved with the help of antidiuretic hormone (ADH), produced by the posterior lobe of the pituitary gland.

Scenario 1: warm environmental conditions

- **Osmoreceptors** in the hypothalamus detect an **increase** in the solute concentration of the blood plasma.
- The **hypothalamus** then produces, by neurosecretion, the hormone **ADH**.
- The ADH is secreted into the posterior lobe of the **pituitary gland**.
- From here it passes into the blood and finally reaches the target organs, the **kidneys**.
- Here it **increases** permeability of:
  (i) the collecting ducts
  (ii) the distal tubules.
- The effect is that more water can be **reabsorbed** back into blood.

The events outlined above give a maximum effect of the countercurrent multiplier. Too much water would be lost by sweating so the water component of the urine must be drastically limited. The resulting urine is therefore low in water content and high in solutes.

Examination tip!

Look out for graphs in questions about kidneys.

- Levels of key substances may be shown.
- If water content down a collecting duct decreases as water content in the medullary region increases
  – then water molecules are crossing the collecting duct
  – sodium and chloride ions have drawn this water from the collecting duct into the medulla by osmosis.

Scenario 2: cold environmental conditions

- **Osmoreceptors** in the hypothalamus detect a **decrease** in the solute concentration of the blood plasma.
- The hypothalamus then produces **less ADH**.
- Less ADH leaves the posterior lobe of the pituitary gland.
- Less ADH reaches the target organs, the kidneys.
- The collecting ducts and the distal tubules are **not so permeable**.
- **Less water** can be **reabsorbed** back.

The urine is of greater volume due to greater water content. No wonder we urinate more in cold weather!

Diuresis

Diuresis is a condition in which excessive amounts of watery urine are produced. In a healthy person this is avoided with the secretion of ADH.

Sometimes people are prescribed a drug equivalent to ADH to cure the symptoms. Reabsorption can take place efficiently so urine at a correct solute concentration is excreted.

4.5 Adaptations to desert ecosystems

After studying this section you should be able to:

- describe how water loss is limited in desert organisms

How do desert organisms limit water loss?

AQA A   M6
AQA B   M6
EDEXCEL M4
OCR   M5
NICCEA   M4

Organisms which successfully live in deserts have structural, physiological and behavioural adaptations. The following organisms display a range of water conserving features.

Kangaroo rat

- Inhaled air accepts heat energy from nasal passages.
- Water vapour from exhaled air condenses on the **cooler** nasal passages and so less water vapour is breathed out.
- It eats plant material, does not drink, but obtains useful **metabolic water** from respiration.
- Remains underground in its burrow for a high proportion of the day, where the humidity is high.
- Less moisture is lost by evaporation from the animal because the **diffusion gradient is reduced**.

> In an examination question you may be given another animal, not the kangaroo rat! Analyse the data, and you will find the principles are similar!

A kangaroo rat

A camel

Camel

> This list of points may be useful in your examination. The temperature range in a day is only applicable in times of water shortage. At an oasis with a plentiful supply of water, camel temperature range in the day is 36°C to 38°C.

- During a period of water deprivation, can routinely have a core temperature from 35°C in the morning to around 42°C in early evening.
- Toleration to the increased temperature gives a period for the camel to cool down slowly by **conduction**, **convection** and **radiation**.
- If cooling was by water evaporation from the animal, then an excessive amount of water would be needed (not possible in desert journeys!).
- The respiration of fat also supplies metabolites, useful to the camel in other processes.
  - Have very long loops of Henlé so that much more water reabsorption can take place.
- The reason for this is that a greater length allows for more Na^+ and Cl^- to be actively transported out of the loop of Henlé, hence greater water retention.

Sample question and model answer

The diagram shows a cell of the inner wall of a renal (Bowman's) capsule. These two structures shown in the diagram are very important in the passage of substances out of the blood into the proximal tubule.

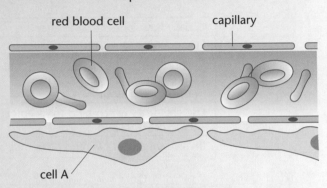

Note the close proximity of cell A to the capillary. This gives a clue as to their function.

(a) Name cell A. [1]

podocyte

(b) Explain how the cells of the inner capsule wall and the capillaries of the glomerulus help in the process of ultrafiltration. [5]

The question shows that five marks are available. Make sure that you give at least five points to gain you marks. Superficial answers fall short of the total.

capillaries lie very close to the inner capsular wall; the capillaries have pores; the podocytes are shaped so that many gaps exist between the capsular wall and capillaries; the resistance to flow caused by tiny pores and gaps contributes to high pressure; only molecules which are small enough are forced through pores so the process is selective.

(c) As glomerular filtrate leaves the renal (Bowman's) capsule it enters the proximal convoluted tubule. The graph below shows the ratio of glucose and urea in the blood plasma and the filtrate through the proximal tubule. A ratio of 1.0 means that the concentration in both plasma and filtrate are the same.

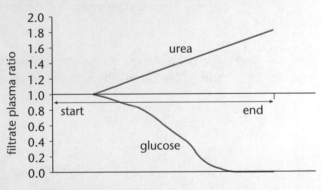

Explain the changes in plasma–filtrate from the beginning to the end of the proximal tubule for:

You may find this difficult. However, you can link the fact that the kidney nephron does reabsorb useful substances but not the waste, urea. Relate this to the graph then your task is possible!

(i) glucose [3]

As the fluid moves along the tubule there is increasingly more glucose in the plasma than the filtrate; this is because glucose is reabsorbed into blood; all glucose is returned to blood before end of proximal tubule.

(ii) urea. [3]

As the fluid moves along the tubule there is increasingly more urea in the filtrate than in the blood plasma; no urea is reabsorbed so it remains in tubule; water is reabsorbed which has the effect of increasing the relative concentration of urea.

Practice examination questions

1 (a) Complete the table below to compare the nervous and endocrine systems. Put a tick in each correct box for the features shown.

	Nervous system	Endocrine system
Usually have longer lasting effects		
Have cells which secrete transmitter molecules		
Cells communicate by substances in the blood plasma		
Use chemicals which bind to receptor sites in cell surface proteins		
Involve the use of Na^+ and K^+ pumps		

[2]

(b) Name the process which keeps the human body temperature and water content of blood regulated. [1]

[Total: 3]

2 A mammal is in hot environmental conditions. Explain the effect of a high quantity of ADH entering the blood from the pituitary gland. [6]

3 (a) The products of transamination have been represented below. Complete the equation

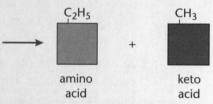

amino   keto
acid   acid   [2]

(b) (i) Where in the human body does transamination take place? [1]
    (ii) Why is transamination necessary in the human body? [2]

[Total: 5]

4 The graph below shows the relative levels of glucose in the blood of two people A and B. One is healthy and the other one is diabetic.

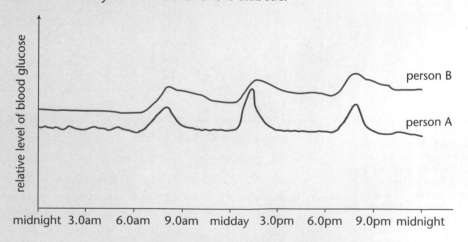

(a) Which person is diabetic? Give evidence from the graph for your answer. [1]

(b) What is the evidence that both of the people do produce insulin? [1]

(c) Where in the body is insulin produced? [2]

[Total: 4]

Chapter 5

Further genetics

The following topics are covered in this chapter:

- *Genes, alleles and chromosomes*
- *Cell division*
- *Inheritance*
- *Applications of genetics*

5.1 Genes, alleles and chromosomes

After studying this section you should be able to:

- *define a range of important genetic terms*
- *understand the origin and range of mutations*

LEARNING SUMMARY

Essential genetic terms

AQA A	M5
AQA B	M4
EDEXCEL	M5
OCR	M5
WJEC	M5
NICCEA	M5

A gene is a section of DNA which controls the production of a protein in an organism. The total effects of all of the genes of an organism are responsible for the characteristics of that organism. Each protein contributes to these characteristics whatever its role, e.g. structural, enzymic or hormonal.

It is necessary to understand the following specialist range of terms used in genetics.

Allele – an alternative form of a gene, always located on the same position along a chromosome.

> E.g. white colour of petals

Check out all of these genetic terms.

- Look carefully at the technique of giving an example with each definition. Often examples help to clarify your answer and are usually accepted by the examiners.
- In examination papers you will need to apply your understanding to **new** situations.
- Genetics has a specialist language which you will need to use.

Dominant allele – if an organism has two different alleles then this is the one which is expressed, often represented by a capital letter.

> E.g. red colour pigment of petals, **R**

Recessive allele – if an organism has two different alleles then this is the one which is **not** expressed, often represented by a lower case letter. Recessive alleles are only expressed when they are not masked by the presence of a dominant allele.

> E.g. white colour pigment of petals, **r**

Homozygous – refers to the fact that in a diploid organism both alleles are the same.

> E.g. **R R** or **r r**

Heterozygous – refers to the fact that in a diploid organism both alleles are different.

Key points from AS

- **The genetic code**
  Revise AS pages 82–84

> E.g. **R r** (petal colour would be expressed as red)

Co-dominance – refers to the fact that occasionally two alleles are equally expressed in the organism.

E.g. **A, B** alleles = **AB** (blood group with antigens A and B)

Polygenic inheritance – where an inherited feature is controlled by two or more genes, along different loci along a chromosome. Results in continuous variation.

E.g. height of a person is controlled by a number of different genes.

> A number of inherited alleles of a range of genes often exhibit continuous variation, e.g. height. Each allele contributes small incremental differences. That is why there are smooth changes in height across a population.

> Remember that both sperms and ova are haploid.

Haploid – refers to a cell which has a single set of chromosomes.
E.g. a nucleus in a human sperm has 23 single chromosomes.

Diploid – refers to a cell which has two sets of chromosomes.
E.g. a nucleus in a human liver cell has 23 pairs of chromosomes.

> In diploid cells one set of chromosomes is from the male parent and one from the female.

Homologous chromosomes – refers to the pairs of chromosomes seen during cell division. These chromosomes lie side by side, each gene at each locus being the same.

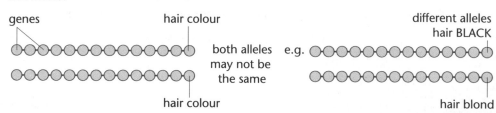

> Often polyploid organisms cannot reproduce sexually but asexually they are successful.

Polyploid – refers to the fact that a cell has three or more sets of chromosomes. This can increase yield.

E.g. cultivated potato plants are **tetraploid**, that is four sets of chromosomes in a cell. (*Tetraploidy is a form of polyploidy*.)

> **phenotype = genotype + environment**
>
> Phenotype includes all alleles which are expressed in an organism. The environment supplies resources and conditions for development. Varying conditions result in an organism developing differently. Identical twins fed different diets will show some differences, e.g. weight.
>
> Environment has considerable effect.

Genotype – refers to all of the genes found in the nuclei of an organism, including both dominant and recessive alleles.

dominant

E.g. A B c d E F g H i   (all alleles are included in a genotype)
    a B C D e f g h I

recessive

Phenotype – refers to only the alleles of an organism which are expressed (the appearance of an organism).

E.g. **A B** c d **E F** g **H** i   only alleles in bold included in a phenotype
    a **B C D** e f g h **I**

Linkage – refers to two or more genes which are located on the same chromosome.

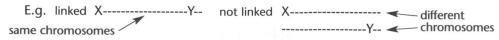

Somatic cell – refers to any cell which is not involved in reproduction.
E.g. liver cell

Autosome – refers to every chromosome apart from the sex chromosomes, X and Y.

Bases can change along DNA and this may cause mutation. One changed base along the coding strand of DNA may have a sequential effect of changing most amino acids along a polypeptide.

before mutation
TTA CCG GCC ATC

after mutation
ATT ACC GGC CAT C

This is addition!

More mutations shown below. Each section of DNA along the chromosomes is shown by organic bases.

Addition

before
TTA CCG GCC ATC

after
CCG TTA CCG GCC ATC

A new triplet has been added. If a triplet is repeated it is also duplication.

Deletion

before
TTA CCG GCC ATC

after
TTA CCG GCC

Inversion

before
TTA CCG GCC ATC

after
TTA CCG GCC **CTA**

CTA codes for new amino acid.

Translocation

before
TTA CCG GCC ATC

after
TTA CCG GCC ATC **CAT**

CAT broke away from another chromosome.

Gene therapy – where the function of defective genes of an organism is rectified by supplying correct DNA material.

E.g. symptoms of the genetic disease, cystic fibrosis, are relieved by applying a 'blast' of corrected genes into alveoli in the lungs. (This is experimental, symptom relief is currently very limited.)

Mutation

Mutation is a change in the DNA of a cell. If the cell affected by mutation is a **somatic cell**, then its effect is **restricted** to the organism itself. If, however, the mutation affects **gametes**, then the genetic change will be inherited by the future population.

> DNA codes for the sequence of amino acids along polypeptides and ultimately the characteristics of an organism. Each amino acid is coded for by a triplet of bases along the coding strand of DNA, e.g. TTA codes for threonine. The change in a triplet base code can result in a new amino acid, e.g. ATT codes for serine. This type of DNA change along a chromosome is known as a **point mutation**. A point mutation involves a change in a small section along a chromosome by **addition**, **deletion** or **inversion**.
>
> If a complete chromosome is added or deleted, this is a **chromosomal mutation**, such as Down's syndrome where a person has an additional chromosome, totalling 47 in each nucleus rather than the usual 46.

KEY POINT

What causes mutations?

All organisms tend to mutate randomly, so different sections of DNA can appear to alter by chance. The appearance of such a random mutation is usually very rare, typically one mutation in many thousands of individuals in a population. The rate can be increased by **mutagens** such as:

- **ionising radiation** – including ultra violet light, X rays and α, β and γ (gamma) rays and neutrons. These forms of radiation tend to dislodge the electrons of atoms and so disrupt the bonding of the DNA which may re-bond in different combinations.
- **chemicals** – including asbestos, tobacco, nitrous oxide, mustard gas and many substances used in industrial processes such as vinyl chloride. Many pesticides are suspected mutagens. Dichlorvos, an insecticide, is a proven mutagen.

Additionally colchicine is a chemical derived from the Autumn crocus, *Colchicium*, which stimulates the development of extra sets of chromosomes.

Are mutations harmful or helpful?

An individual mutation may be either harmful or helpful. When tobacco is smoked this can increase the rate of mutation in some somatic cells. The DNA disruption can result in the formation of a cell which divides uncontrollably and causes the disruption of normal body processes. This is **cancer**, and can be lethal.

Chrysanthemum plants have a high rate of mutation. A chrysanthemum grower will often see a new colour flower on a plant, e.g. a plant with red flowers could develop a side shoot which has a different colour, such as bronze. Most modern chrysanthemums appeared in this way, production being by asexual techniques.

Some mutated human genes have through evolution been successful. Many successful mutations contributed to the greater size of cerebrum, proportionally than other primates.

5.2 Cell division

After studying this section you should be able to:

- *compare the main features of mitosis and meiosis*
- *describe and explain the process of meiosis*
- *understand the consequences of chiasmata (crossing over)*

LEARNING SUMMARY

Why are there two types of cell division?

AQA A	M5
AQA B	M4
EDEXCEL	M5
OCR	M5
WJEC	M5
NICCEA	M5

Examiner's tip

At AS Level you learned the names of the stages in sequence. The stages of meiosis use the same names, in the same order but there are two nuclear divisions this time!

Each type of cell division has a different purpose.

Mitosis

There are occasions when it is necessary to **replicate** cells, e.g. in growth and repair. This is the role of **mitosis**. It produces a clonal line of cells. Each cell divides to form **2 diploid**, daughter cells, identical in every way.

Meiosis

This is needed in gamete formation. In human cells a body (somatic) cell has 46 chromosomes. If each gamete contained 46 chromosomes then the zygote produced at fertilisation would have 92 chromosomes. This would be lethal! Meiosis is also called **reduction division** because the gametes produced are haploid. In human gametes the haploid chromosome number is 23. Each cell divides to form **4 haploid**, daughter cells. Every daughter cell is different to the parent cell and each other.

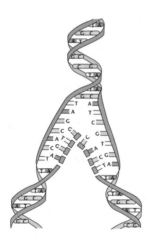

Meiosis: the process explained

The preparation of a cell prior to meiotic division is during **interphase**. During this pre-stage each double strand of DNA replicates to produce **two** exact copies of itself. This also takes place in exactly the same way before mitosis takes place. After interphase, when the cell division commences, major differences occur.

In meiosis during the first stage, **prophase 1**, a fundamentally important event takes place, where chromatids **cross over**. Each crossover is termed a **chiasma**.

The mechanism of cross overs (chiasmata)

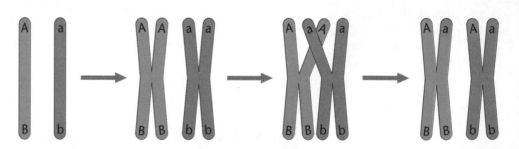

A represents an allele dominant to **a**, a recessive allele.

B represents an allele dominant to **b**, a recessive allele.

In humans, with **many chiasmata** taking place along **all 23 pairs of chromosomes** every cell at the completion of meiosis is genetically different.

Chiasmata result in **different allele combinations**!

KEY POINT

Key points from AS

- **Cell division**
  Revise AS pages 85–86

The process of meiosis

In the division of a human cell by meiosis there are 23 pairs of chromosomes in the parent cell. If all 46 chromosomes were represented in diagrams then there would be confusion. In these diagrams only 2 pairs of chromosomes are shown, but remember there are 21 other pairs! One homologous pair is shown in two colours to show the consequence of crossovers.

early prophase I

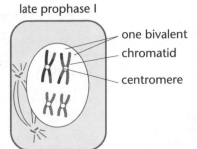

one homologous pair of chromosomes

chromosome

centriole

each chromosome forms 2 chromatids 2 centrioles begin to move forming a spindle

late prophase I

one bivalent

chromatid

centromere

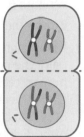

the chromatids have crossed over and exchanged DNA at 2 positions

metaphase I

bivalents form homologous chromosomes lie parallel to each other along the equator

anaphase I

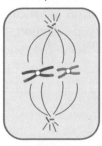

corresponding bivalents are pulled by spindle fibres towards poles

telophase I

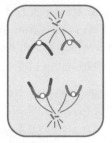

cell constricts at equator to form daughter cells

prophase II

EACH of these cells will divide to form 2 daughter cells

metaphase II

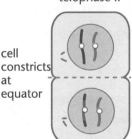

single bivalents lie across the equator

anaphase II

spindle fibres contract to pull centromere apart. Single chromosomes dragged to the poles

early telophase II

cell constricts at equator

2 daughter cells produced for each cell form first division = 4 daughter cells

The significance of meiosis

Diagram to show the single chromosomes produced as a result of two crossovers.

end of first meiotic division

'sister' chromatids still attached

end of second meiotic division

'sister' chromatids have parted from centromere

The combination of each of these chromosomes with others, results in further genetic variation

In an examination you will need to understand the consequence of many crossovers. Crossovers are a source of genetic variation.

A represents an allele dominant to **a**, a recessive allele.

B represents an allele dominant to **b**, a recessive allele.

- Many more than two crossovers can take place between each homologous pair! The presence of 23 homologous pairs of chromosomes in a diploid human cell result in a lot of crossovers.
- Once the chromatids finally separate in **anaphase II**, each moves with 22 others to a pole to produce a daughter cell.
- After division, four different chromatids are produced from each homologous pair (see above).

1 from 4 chromatids combine with 1 from another 4 chromatids. These combinations give **16 possibilities**. Add the combination of another 1 from 4 chromatids and there are **64 possibilities**. Another 1 from 4 is added .... and another .... to include all 23 pairs. This gives millions of combinations. No wonder we all look different!

> **KEY POINT**
>
> What determines which chromatid from the four of one homologous pair is grouped with chromatids from the other homologous pairs?
>
> The answer is '**chance**' and the combination of the 23 single chromosomes dragged through the cytoplasm by the spindle fibres, is known as **independent assortment**.

Every gamete produced by meiosis is genetically different. However, there are two sexes. This means that in sexual reproduction, the fact that there are two different gametes which combine their alleles in the zygote, this gives another major source of variation.

Progress check

The diagram below shows a stage in cell division.

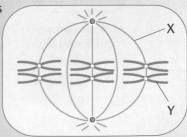

(a) Name parts X and Y.

(b) Which type of cell division is shown? Give a reason for your answer.

(b) Metaphase I of meiosis. Bivalents line up in twos along the equator whereas in mitosis they lie singly.

(a) X = spindle fibre Y = chromatid or bivalent

5.3 Inheritance

After studying this section you should be able to:

- *understand Mendel's laws of inheritance*
- *understand the principles of monohybrid inheritance and dihybrid inheritance*
- *understand the terms parental DNA, recombinant DNA and co-dominance*
- *describe the principle of sex determination*
- *use the Hardy–Weinberg Principle to predict the numbers of future genotypes*

LEARNING SUMMARY

Mendel and the laws of inheritance

AQA A	M5
AQA B	M4
EDEXCEL	M5
OCR	M5
WJEC	M5
NICCEA	M5

Gregor Mendel was the monk who gave us our understanding of genetics. He worked with organisms such as pea plants to work out genetic relationships.

> **Mendel's first law indicates that:**
> - each character of a diploid organism is controlled by a pair of alleles
> - from this pair of alleles only one can be represented in a gamete.
>
> **KEY POINT**

Always show your working out of a genetical relationship in a logical way, just like solving a mathematics problem.

Monohybrid inheritance

Mendel found that when homozygous pea plants were crossed, a predictable ratio resulted. The cross below shows Mendel's principle.

pea plants   pea plants
T = TALL (dominant)   t = dwarf (recessive)

TOP TIP ALERT!

If you have to choose the symbols to explain genetics, then use something like **N** and **n**. Here the upper and lower cases are very different. **S** and **s** are corrupted as you write quickly and may be confused by the examiner awarding your marks.

A homozygous TALL plant was crossed with a homozygous recessive plant

$$TT \quad x \quad tt$$

gametes   (T)   (T)   (t)   (t)

F_1 generation   Tt

All offspring 100% TALL, and heterozygous.

Heterozygous plants were crossed

$$Tt \quad x \quad Tt$$

gametes   (T)   (t)   (T)   (t)

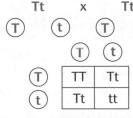

2 × 2 punnet square to work out different genotypes

F_2 generation   3 TALL : 1 DWARF

In examinations you may have to work out a probability. 3 : 1 is the same as a 1 in 4 chance. Remember only large numbers would confirm the ratio.

In making this cross Mendel investigated **one** gene only. The height differences of the plants was due to the different alleles. Mendel kept all environmental conditions the same for all seedlings as they developed. The 3:1 ratio of tall to short plants only holds true for large numbers of offspring.

Dihybrid Inheritance

Mendel again investigated using pea plants to work out the genetic relationship between plants for genes at different loci (positions) on chromosomes.

> **Mendel's second law of independent assortment indicates that:**
> - either of a pair of alleles, say **A** and **a**
> - can combine with either of another pair, say **B** and **b**.
>
> **KEY POINT**

Be ready to cross two organisms from a punnet square, e.g.

RrYY x rrYy

Do not expect a 9:3:3:1 ratio!

The cross below shows Mendel's dihybrid principle.

pea plants	*pea plants*
R = round seeds (dominant)	r = wrinkled seeds (recessive)
Y = yellow seeds (dominant)	y = green seeds (recessive)

A homozygous dominant plant with yellow, round seeds was crossed with a homozygous recessive plant with green, wrinkled seeds.

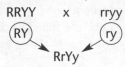

$$RRYY \quad x \quad rryy$$

gametes   (RY)   (ry)

F_1 generation   RrYy

seeds 100% round, yellow, and heterozygous

Heterozygous plants from F_1 generation were crossed.

RrYy   x   RrYy

gametes   (RY) (Ry) (rY) (ry)   (RY) (Ry) (rY) (ry)

F_1 generation

	RY	Ry	rY	ry
RY	RRYY	RRYy	RrYY	RrYy
Ry	RRYy	RRyy	RrYy	Rryy
rY	RrYY	RrYy	rrYY	rrYy
ry	RrYy	RRyy	rrYy	rryy

4 × 4 punnet square to work out genotypes

F_1 generation

9	:	3	:	3	:	1	ratio
round yellow		round green		wrinkled yellow		wrinkled green	

The 9 : 3 : 3 : 1 ratio only holds true for large numbers of offspring.

The principles above can be applied to any dihybrid example. The F_1 generation is so predictable that many varieties of commercial crop are grown from F_1 generation seeds, known as F_1 hybrids, e.g. Brussels sprout, variety 'Peer Gynt'. A uniform crop of high yield.

Linkage

Linkage can occur on autosomes and sex chromosomes.

Each chromosome consists of a sequence of genes. All genes along a chromosome are **linked** because they are part of the same chromosome. Most chromosomes have between 500 and 1000 genes in a linear sequence. These genes are linked.

What is the significance of linkage?

We are able to make predictions about the proportion of future offspring when we know the genotype of parents, like the 9:3:3:1 ratio for dihybrid inheritance. This is only true if the pair of contrasting genes are **on different chromosomes**. Consider these two alternatives:

A dominant, **a** recessive; **B** dominant, **b** recessive

loci (positions) of genes

AaBb x AaBb

Not linked. This cross would produce 9:3:3:1 proportion in offspring

AaBb x AaBb

Linked. This cross would be unlikely to produce a 9:3:3:1 proportion in offspring

The more crossovers there are, the greater the chance that the four different gene combinations will be produced in each parental genotype. They could produce a 9:3:3:1 proportion in cross 1. However, if the genes are closer along the chromosome then the proportions deviate significantly from this pattern. Genes **adjacent** to each other tend to be **inherited together**, because the chance of them being parted is very low.

When two genes, e.g. A,a and B,b are on **different chromosomes** then their inheritance together is **not affected by crossovers**. Either of one pair **can** be inherited with either of the other pair. The relationship changes when the genes are along the same chromosome. Crossovers are affected! Alleles along the same gene locus can be swapped from one chromatid to another.

Consider these alternatives for linked genes, where a homozygous dominant genotype is crossed with a homozygous recessive.

Crossover 1 (AABB x aabb)

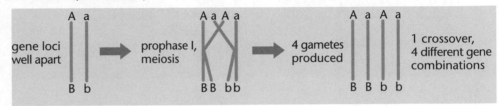

Crossover 2 (AABB x aabb)

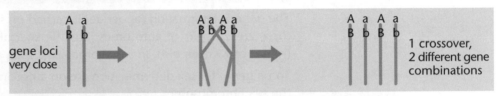

Linkage and probability

Consider these crosses

Cross one

R = red petals (dominant)   r = blue (recessive)
L = long stems (dominant)   l = short stems (recessive)
   homozygous red petals   homozygous blue petals
   long stemmed   short stemmed

                    RRLL   x   rrll

gametes   R   R   r   r
               L   L   l   l

F₁ generation   RrLl   100% heterozygous
                                       red petals, long stemmed

The young organisms are often termed progeny or offspring.

Cross two

Heterozygous   Heterozygous
red petals   red petals
long stemmed   long stemmed

            R r L l   x   R r L l

gametes   R   r   R   r
            L · l   L   l

F₂ generation

	RL	rl
RL	RRLL red long	RrLl red long
rl	RrLl red long	rrll blue short

probability is: red petals 3 : 1 blue petals
                       long stems   short stems
actual numbers   610   202

(That is almost the one in four chance!)

Examiner's tip

Do not fall into the 'linkage' trap. Here the 9:3:3:1 ratio is replaced by a 3:1 ratio due to genes being adjacent on the chromosome. The net result is that **RL** and **rl** tend to be inherited together as a package. Do not dive into questions like this. Keep alert!

This example shows the consequence of **very close linkage**. In this genuine example the genes were so close that the RL and rl combinations were never parted by crossovers. No Rl or rL allele combinations were evident. So the classic RrLl x RrLl ratio of 9 : 3 : 3 : 1 was not possible. Instead a 3 : 1 ratio was produced. This is **not** monohybrid inheritance!

Progress check

(a) List the gametes for the following dihybrid cross.
    (The genes are not linked.)
        Ddee x DDEe

(b) Show the genotypes of the progeny.

(b) DDEe, Ddee, DDee, Ddee
(a) De de   DE De

Sex determination

The genetic information for gender is carried on specific chromosomes. In humans there are 22 pairs of autosomes plus the special sex determining pair, either **XY** (**male**) or **XX** (**female**). In some organisms such as birds this is reversed.

Some genes for sex determination are on autosomes but are activated by genes on the sex chromosomes.

Sperms can carry an X or Y chromosome, whereas an egg carries only an X chromosome.

> The genetic cross shown should not give you any problems at A2 Level. However, look out for the combination of another factor which will increase difficulty.

genotype	XX	x	XY
	female		male

gametes	Ⓧ	Ⓧ	Ⓧ	Ⓨ
		Ⓧ	Ⓧ	

offspring		X	X
Ⓧ		XX	XX
Ⓨ		XY	XY

probability	1	:	1
	male		female

This shows how 50 : 50 males to females are produced.

Sex linkage

Look more closely at the structure of the X and Y chromosomes.

> Why do they not look like X and Y? Only when the cells are dividing, do they take the X,Y shape, after chromatid formation.

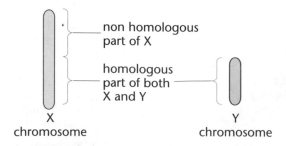

non homologous part of X

homologous part of both X and Y

X chromosome   Y chromosome

Homologous part of the sex chromosomes

• Has the same genes in both sexes.
• Each gene can be represented by the same or different alleles at each locus.

Non-homologous part of the sex chromosomes

- This means that the X chromosome has genes in this area, whereas the Y chromosome, being shorter, has no corresponding genes.
- Genes in this area of the X chromosome are always expressed, because there is no potential of a dominant allele to mask them.
- There are some notable genes found on the non-homologous part, e.g. haemophilia trait, and colour blindness trait.

The sex chromosomes, X and Y, carry genes other than those involved in sex determination. Examples of such genes are:

- a gene which controls the ability to detect red and green colours
- a gene which controls blood clotting; i.e. responsible for the production of factor VIII vital in the clotting process.

The loci of both genes are on the non-homologous part of chromosome X.

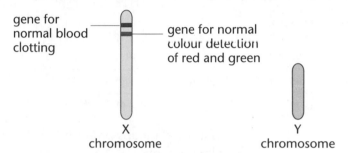

What is the effect of sex linked genes?

The fact that these genes are linked to the X chromosome has no significant effect when the **genes perform their functions correctly**. There are consequences, however, if the genes fail. This can be illustrated by a consideration of **red-green colour blindness**. When a gene is carried on a sex chromosome, the usual way to show this is by X^R.

R = normal colour vision (dominant)   r = red-green colour blindness (recessive)

You can see the four possible genotypes. A female needs two r alleles, a male just need one!

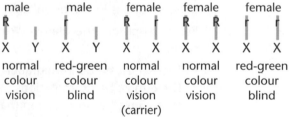

The genetic diagram shows that a female needs two recessive alleles (one from each parent!) to be colour blind. A male has only one gene at this locus, so one recessive allele is enough to give colour blindness.

The colour blindness gene is rare, so the chances of being a colour blind female are very low. Consider these crosses.

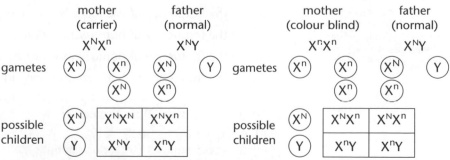

1 in 4 of the children is red-green colour blind. 50% of boys are colour blind, but no girls.

All boys are red-green colour blind. No girls are colour blind.

Co-dominance

Remember that for co-dominance there is no dominance. Both alleles are equally expressed.

This term is given when each of two *different* alleles of a gene are expressed in the phenotype of an organism. In humans there are two co-dominant alleles. These alleles produce the antigens in blood which are responsible for our blood groups.

Consider these crosses

Blood group	Genotypes
A	$I^A I^A$, $I^A I^O$
B	$I^B I^B$, $I^B I^O$
AB	$I^A I^B$
O	$I^O I^O$

The allele for production of:

A antigen in blood = I^A

B antigen in blood = I^B

No antigen in blood = I^O

Examiner's tip

Look out for more examples of co-dominance in examination questions, e.g. in shorthorn cattle, R = red and W = white.

Where they occur in the phenotype together they produce a dappled intermediary colour known as roan.

Mother × Father

$I^B I^O$ $I^A I^O$

gametes I^B I^O I^A I^O

children

	I^B	I^O
I^A	$I^A I^B$	$I^A I^O$
I^O	$I^B I^O$	$I^O I^O$

All blood groups produced by this cross.

Mother × Father

$I^A I^B$ $I^A I^O$

gametes I^A I^B I^A I^O

children

	I^A	I^B
I^A	$I^A I^A$	$I^A I^B$
I^O	$I^A I^O$	$I^B I^O$

There must be a I^O from both parents to produce an O blood group.

In this instance there are two co-dominant alleles, I^A and I^B. When inherited together they are both expressed in the phenotype. Group O blood does not have any antigen.

Hardy–Weinberg Principle

The application of this principle allows us to **predict numbers of expected genotypes** in a population in the future. The principle tracks the proportion of two different alleles in the population.

Before applying the Hardy–Weinberg principle the following criteria must be satisfied.

- There must be no immigration and no emigration.
- There must be no mutations.
- There must be no selection (natural or artificial).
- There must be true random mating.
- All genotypes must be equally fertile.

Once the above criteria are satisfied then **gene frequencies remain constant**.

A **gene pool** consists of all genes and their alleles, which are part of the reproductive cells of an organism. Only genes that are in cells that **can be passed on** are part of the gene pool.

Hardy–Weinberg Principle: the terms identified

p   = the frequency of the dominant allele in the population

q   = the frequency of the recessive allele in the population

p^2 = the frequency of homozygous dominant individuals

q^2 = the frequency of homozygous recessive individuals

2pq = the frequency of heterozygous individuals

The principle is based on two equations:

(i) $p + q = 1$ (gene pool)

(ii) $p^2 + 2pq + q^2 = 1$ (total population)

KEY POINT

Applying the Hardy–Weinberg Principle

A population of *Cepaea nemoralis* (land snail) lived in a field. In a survey there were 1400 pink-shelled snails and 600 were yellow. There were two alleles for shell colour.

y = yellow shell (recessive) Y = pink shell (dominant). Snails with pink shells can be YY or Yy. Snails with yellow shells can be yy only

phenotype	pink	yellow
genotype	YY Yy	yy

This part of the calculation is to find the frequency of the recessive and dominant alleles in the population.

$$q^2 = \frac{600}{2000}$$
$$= 0.3$$
$$q = \sqrt{0.3} = 0.55$$

But $p + q = 1$
$$p = 1 - 0.55$$
$$= 0.45$$

So $p^2 = 0.20$

This part of the calculation is to find the frequency of homozygous and heterozygous snails in the population.

But $p^2 + 2pq + q^2 = 1$
$$0.20 + 0.5 + 0.3 = 1$$
$$YY \quad Yy \quad yy$$

Points to note

* These proportions can be applied to the snail populations say, 10 years in the future.
* If there were 24 000 snails in the population then the relative numbers would be:
  YY $0.2 \times 24\,000 = 4800$
  Yy $0.5 \times 24\,000 = 12\,000$
  yy $0.3 \times 24\,000 = 7200$
* Remember that the five criteria must be satisfied if the relationship is to hold true.
* It is not possible to see which snails are homozygous dominant and which are heterozygous. They all look the same, pink! Hardy–Weinberg informs us, statistically, of those proportions.

It is also possible to apply the Hardy–Weinberg principle to a co-dominant pair of alleles. P and q are calculated by exactly the same method.

Examiner's tip

Always use the $p + q = 1$ equation to calculate the frequency of alleles if you are given suitable data, e.g. 'out of 400 diploid organisms in a population there were 40 homozygous recessive individuals.' 40 organisms have 80 recessive alleles.

$$q = \frac{80}{800}$$
$$= 0.1$$

From this figure you can calculate the others.

Progress check

What is the probability of a colour blind male and carrier female producing:

(a) a boy with normal colour vision
(b) a colour blind girl? Show your working.

(a) 1 in 4 (b) 1 in 4

mother (carrier)	father (colour blind)	
$X^N X^n$	$X^n Y$	

	X^N	X^n
X^n	$X^N X^n$	$X^n X^n$
Y	$X^N Y$	$X^n Y$

Epistasis

This involves two different genes which affect each other. A form of epistasis can be explained by referring to the sweet pea plant. *Lathyrus odoratus* is a white flowered sweet pea. When crossed, two white parent plants can produce white and purple flowers. This can be explained as follows:

Dominant epistasis also exists. In this instance a dominant allele can **inhibit** another, e.g. in the land snail **A a** dominant allele inhibits **B, b** alleles responsible for banding on the shell.

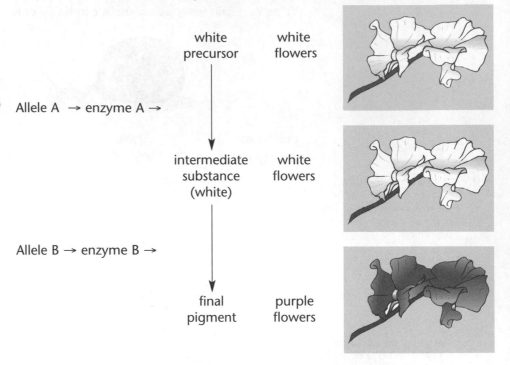

Allele A → enzyme A →

white precursor   white flowers

intermediate substance (white)   white flowers

Allele B → enzyme B →

final pigment   purple flowers

Both alleles A and B are needed to code for their respective enzymes if purple sweet pea flowers are to be produced. The alleles to consider are:

 A (dominant)   a (recessive)   B (dominant)   b (recessive)

Genotypes	aabb	aaBb	aaBB	Aabb	AaBB	AaBb	AAbb	AABb	AABB
Phenotypes	white	white	white	white	purple	purple	white	purple	purple

Without the combined effects of both A and B alleles then the flowers are white. The reliance of one gene on another is an example of **epistasis**.

How is it possible for two white flowered plants to be crossed to give purple offspring?

Questions about epistasis usually give some data which you will need to analyse. The organisms may not be sweet pea plants but the principles remain the same.

	white		white
genotype	AAbb	x	aaBB
gametes	Ab Ab		aB aB
F₁ generation		AaBb	

100% purple flowered plants from white flowered parents.

Check out the above genotypes to find two more genotypes of white flowered plants which could be crossed to give purple offspring.

Chi-squared: a statistical test

When doing scientific investigations we need to know if our results are significant or due to chance. We should not, for example, conclude that a new genetic ratio we have found represents a significant pattern for a particular cross. The x^2 (**chi-squared**) test helps us to check out the difference between **expected** results and **actual** results. We can then state the probability that any differences between expected and actual results are due to chance or have significance.

> Remember in **co-dominance** both alleles are expressed in the phenotype.

$$x^2 = \sum \frac{d^2}{x}$$

d = difference between actual and expected results
x = expected results
Σ = the sum of

Consider this example

Dianthus (campion) has flowers of three different colours, red, pink and white. Two pink flowered plants were crossed and the collected seeds grown to the flowering stage.

R = red   r = white (both alleles are co-dominant)

genotypes	Rr x Rr
gametes	R r R r
F_1 generation	R r

> In an examination you may be given another term for 'actual'. It may be observed, but it means the same!

	R	r
R	RR	Rr
r	Rr	rr

white 0.25
pink 0.5
red 0.25

numbers	RR = red flowers	Rr = pink flowers	rr = white flowers
actual	34	84	42
expected	40	80	40

$$x^2 = \frac{(40-34)^2}{40} + \frac{(80-84)^2}{80} + \frac{(40-42)^2}{40}$$

$$= \quad 0.9 \quad + \quad 0.2 \quad + \quad 0.1$$

$$= \quad 1.2$$

The next stage is to assess the degrees of freedom for this investigation. This value is always one less than the number of classes of results. In this case there are three classes, i.e. red, pink and white.

Degrees of freedom = (3 − 1) = 2

Now check the x^2 value against the table.

Degrees of freedom	No of classes	x^2							
1	2	0.00	0.10	0.45	1.32	2.71	3.84	5.41	6.64
2	3	0.02	0.58	1.39	2.77	4.61	5.99	7.82	9.21
Probability that deviation is due to chance alone		0.99 (99%)	0.75 (75%)	0.50 (50%)	0.25 (25%)	0.10 (10%)	0.05 (5%)	0.02 (2%)	0.01 (1%)

> If you are given a x^2 question in an examination you will be given a data table. A mark may be given for degrees of freedom. Remember, **10** classes of results would give **9** degrees of freedom.

What do you do with the x^2 value?

- Go to the 2 degrees of freedom line (highlighted).
- Find the nearest figures to 1.2, which comes between the 70% and 50% columns.
- A x^2 value of 1.2 shows that it is at least 70% probable that the result is by chance alone.
- The difference of this result against the expected is **not significant**.

For a **significant difference** the value should fall between (1–5)% columns.

5.4 Applications of genetics

After studying this section you should be able to:

- *understand the principles of artificial selection*
- *describe the technique of micropropagation*
- *outline the inheritance and symptoms of cystic fibrosis, Huntington's chorea and Down's syndrome*
- *understand the need for genetic screening and counselling*
- *understand the need for genetic compatibility in transplant surgery*
- *understand potential applications which may result from the human genome project*

LEARNING SUMMARY

Artificial selection

AQA A	M5
AQA B	M4
EDEXCEL	M5
OCR	M5
WJEC	M5
NICCEA	M5

Compare artificial selection with natural selection (see p.102).

The two processes have similarities but in natural selection it is change of the environment which is the selective agent.

Artificial selection is not the only way to improve animals and plants. Genetic modification is another method. A variety of soya bean plants now have resistance to selective herbicide.

Can you suggest four excellent features offered by this new variety?

This is *selective* breeding to improve specific domesticated animals and crop plants. Important points are:

- people are the **selective agents** and choose the parent organisms which will breed
- the organisms are chosen because they have **desired characteristics**
- the aim is to incorporate the desired characteristics from both organisms in their offspring
- the offspring must be **assessed** to find out if they have the desired combination of improvements (there is **no guarantee** that a cross will be successful!)
- offspring which have suitable improvements are used for breeding, the others are deleted from the gene pool (not allowed to breed).

Most modern crops have been produced by artificial selection. The Brussels sprout variety below was produced in this way. Many trials were carried out before the new variety was offered for sale.

Brilliant NEW FOR 2001
F1 Hybrid A brand new early cropping variety which produces dense, dark green buttons of excellent quality in September and October. Suitable for a wide range of soil types it also has a high resistance to powdery mildew and ring spot. Good for freezing. **2152** *pkt* **£2.10**

All modern racehorses have been artificially selected. Champion thoroughbred horses are selected for breeding on the basis of success in races. Only the best racehorses are actually entered in races. The fastest horses at various distances win races, and the right to breed. Continual improvement results as the gene pool is consistently strengthened.

Key points from AS

- **Genetically modified organisms**
  Revise AS page 90

Genetic disorders

AQA A	M5
AQA B	M4
EDEXCEL	M5
OCR	M5
WJEC	M5
NICCEA	M5

There are many genetic diseases in the human population. Modern medical practices reduce the effects of certain symptoms. Serious diseases which can be life threatening are treated, e.g. haemophiliacs are supplied with factor VIII to allow blood clotting to take place.

Survival of people with a severe genetic deficiency maintains the frequency of defective alleles in the population. Sufferers and carriers who breed, may increase the frequency of the allele in the human gene pool.

Some important genetic disorders

Detection:
- by symptoms shown
- prediction can be made for further children
- e.g. if two people who do not suffer from the condition have a child who is cystic, any further children have a 1 in 4 chance of having the condition.

Cystic fibrosis

- In healthy people a gene codes for a protein which functions as a Cl^- pump in epithelial cells.
- The outward movement of Cl^- ions is accompanied by water, effectively lubricating the outside of these lining cells.
- The **cystic fibrosis trait** is carried by a single, **recessive allele**.
- The disorder is only expressed in the homozygous condition (1 in 2000 people).
- In sufferers the **Cl^- pump protein** does not function correctly.
- The fluid is **more viscous** and moves in a sluggish movement over the epithelial surfaces.
- Adverse effects are evident in the pancreas–duodenum area where there is **inhibited movement** of substances through the alimentary canal.
- Serious effects are found in the lungs where the epithelial cells secrete **thick sticky mucus**.
- The mucus **inhibits breathing** seriously, so must be removed each day to relieve symptoms.

Detection:
- by symptoms shown
- since the allele is dominant even heterozygous people have the disease
- where one person has the disease and the other does not, the chance of a child being born with the condition is 1 in 2.

Huntington's chorea

- The disease is carried by a **dominant allele**.
- Symptoms observed are increasing **involuntary movements** and **mental deterioration**.
- 1 in 100 000 carry the allele and, since it is **dominant**, have the disease.
- Often the symptoms are not shown until the post 40s so that the genes can be passed on before the disease is detected.

Detection:
- before birth by **amniocentesis** or **chorionic villi sampling**
- when detected – in mid-pregnancy, test informs of a certain prediction that child will have Down's syndrome.

Down's syndrome

- This takes place as a consequence of the **chromosomes segregating incorrectly during meiosis**.
- The result is that one gamete carries 22 chromosomes whereas the other carries **24 chromosomes**.
- This is known as **non-disjunction**.
- The gamete carrying 24 chromosomes may then fuse with a normal one carrying 23 to form a zygotic cell with **47 chromosomes**.
- The female gametes can be affected in this way, the incidence increasing above the age of thirty five.
- The chromosome defect can be detected; chromosome numbers are checked out by experienced personnel so that early warning can be given.

Screening

This is checking out the population for a variety of diseases by using screening techniques.

Amniocentesis

The process takes place during mid-pregnancy.

- Some fetal cells detach and become suspended in the amniotic fluid.
- A syringe is inserted through the uterus wall into the amniotic fluid and a sample is removed.
- Some fluid is placed into a Petri dish where the fetal cells grow on a nutrient medium.
- The number of chromosomes in the nuclei is counted.
- 47 chromosomes per nucleus gives an early warning of Down's syndrome.

Chorionic villi sampling

This process also takes place during mid-pregnancy.

- Placental cells arise from the zygote, so the fetus and the placenta have identical nuclei.
- Some chorionic villus cells are removed from the placenta.
- Again the number of chromosomes in the nuclei is counted.

Both of the techniques shown are accompanied by use of an ultrasound scanner. This creates a picture of the fetus in the uterus so that damage can be avoided.

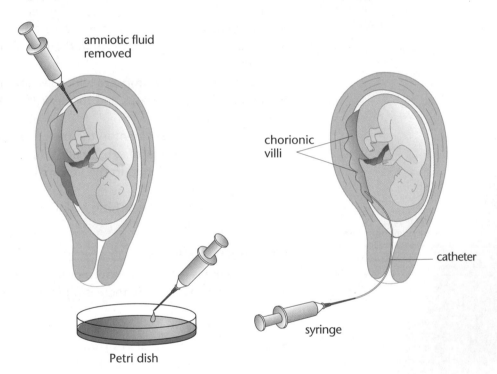

amniotic fluid removed

chorionic villi

catheter

syringe

Petri dish

Genetic counselling

It is useful for potential parents to know the probability of a defective allele being expressed in their future children.

If the genetic pattern of a disease is tracked through a family tree then useful information can be given to potential parents. The probability of having healthy or affected children can be considered and a decision made.

Histocompatibility

This is shown by tissues which can be **transplanted** into a person and refers to the fact they are **not rejected** by the immune system. They have histocompatibility!

Cells have a range of protein molecules on their cell surface membranes. In a person these proteins help to identify a cell as 'self' so that the immune system will not be triggered. All of these proteins are coded for by a person's DNA. Different people have different genes for these histocompatible proteins. The best person to donate an organ would be an identical twin to the recipient! This is rarely possible, so tissue-matching histocompatible cell surface proteins is extremely important. The closer the match, the less will be the intensity of the immune response.

Even with the achievement of a reasonable tissue match it is wise to use **immunosuppressive drugs**. These inhibit the action of the white blood cells and enable the transplant to establish.

The human genome project

This is an analysis of the complete human genetic make-up and will ultimately map the organic base sequences of the nucleotides along our DNA.

A brief history

- 1977 Sanger devised DNA base sequencing.
- 1986 Human genome project initiated in USA and UK.
- 1996 30 000 genes mapped.
- 1999 one billion bases mapped including all of chromosome 22.
- 2000 chromosome 21 mapped with the human genome almost complete.
- 2001 human genome mapping complete.

Some important points

- The genome project will sequence the complete set of over 100 000 genes.
- Only around 5% of the base pairs along the DNA actually result in the expression of characteristics. These DNA sequences are known as **exons**.
- 95% of DNA base sequences do not appear to be involved in the expression of characteristics and are known as **introns**.
- Introns do not outwardly seem to be responsible for characteristics. It is likely that they may be regulatory, perhaps in multiple gene role.

Single nucleotide polymorphisms (SNPs)

Around 99.9% of human DNA is the same in all individuals. Merely 0.1% is different! The different sequences in individuals can be the result of **single nucleotide polymorphism**. One base difference from one individual to another at a site may have no difference. Up to a maximum of six different codons can code for one amino acid. An SNP will not necessarily have any effect.

Some SNPs do change a protein significantly. Such changes may result in genetic disease, resistance or susceptibility to disease.

How can the mapping of SNPs be useful?

- The mapping of SNPs along chromosomes signpost where base differences exist.
- Across the gene pool a pattern of SNP positions will be evident.
- There may be a high frequency of common SNPs found in the DNA of people with a specific disease.
- This highlights interesting sites for future research and will help to find answers to genetic problems.

Benefits obtained from the human genome project

Ultimately the human genome data will be instrumental in the development of drugs to treat genetic disease. Additionally, by analysis of parental DNA, it will be possible to give the probability of the development of a specific disease or susceptibility to it, in offspring. Fetal DNA, obtained through amniocentesis or by chorionic villi sampling, will give genetic information about an individual child. Genetic counsellors will have more information about an individual than ever before. Companies will be able to produce 'designer drugs' to alleviate the problems which originate in our DNA molecules. Soon the race will begin to produce the first crop of drugs to treat or even cure serious genetic diseases.

Look to the media for progress updates!

Side panel

AQA A	M5
AQA B	M4
EDEXCEL	M5
OCR	M5
WJEC	M5
NICCEA	M5

Effects of single nucleotide polymorphism

Example
5 base sequences from five people →

GTATAGCCGCAT 1
GTATAGCCGCAT 1
GTATAGCCGCAT 1
GTATAGCCGCCT 2
GTATAGCCGCCT 2

Version 1 = ●
Version 2 = ●

Proportion of the SNP in healthy members of population:

Proportion of the SNP in diseased members of population:

A greater incidence of an SNP in people with a disease may point to a cause.

Sample questions and model answers

1

(a) Explain the difference between sex linkage and autosomal linkage. [2]

sex linkage – genes are located on a sex chromosome

autosomal linkage – genes are located on one of the other 44 chromosomes

(b) The diagram below shows part of a family tree where some of the people have haemophilia.

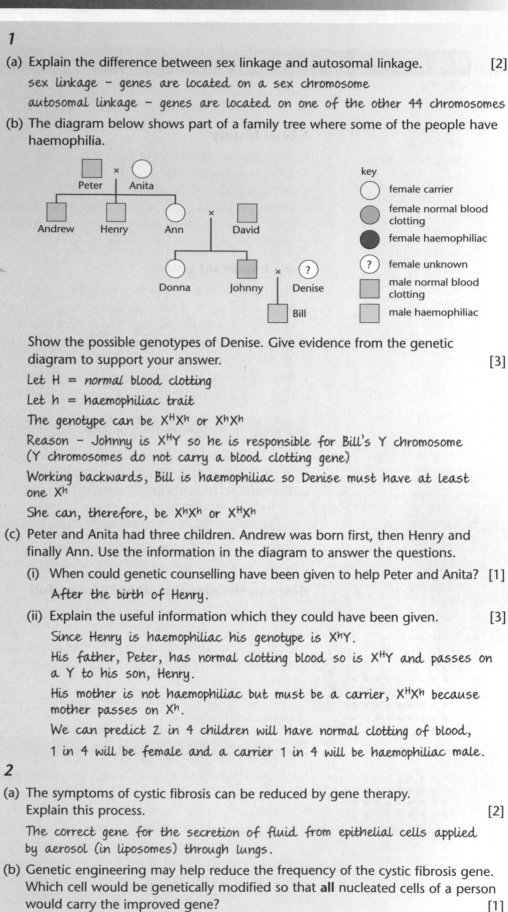

This type of question is a challenge! Note the key for the symbols and then apply them to the family tree. Think logically and work up and down the diagram. In your 'live' examination write on the diagram to help you work out each individual genotype asked in the question. If there are a range of possible genotypes they may be helpful.

Show the possible genotypes of Denise. Give evidence from the genetic diagram to support your answer. [3]

Let H = normal blood clotting

Let h = haemophiliac trait

The genotype can be $X^H X^h$ or $X^h X^h$

Reason – Johnny is $X^H Y$ so he is responsible for Bill's Y chromosome (Y chromosomes do not carry a blood clotting gene)

Working backwards, Bill is haemophiliac so Denise must have at least one X^h

She can, therefore, be $X^h X^h$ or $X^H X^h$

(c) Peter and Anita had three children. Andrew was born first, then Henry and finally Ann. Use the information in the diagram to answer the questions.

(i) When could genetic counselling have been given to help Peter and Anita? [1]

After the birth of Henry.

(ii) Explain the useful information which they could have been given. [3]

Since Henry is haemophiliac his genotype is $X^h Y$.

His father, Peter, has normal clotting blood so is $X^H Y$ and passes on a Y to his son, Henry.

His mother is not haemophiliac but must be a carrier, $X^H X^h$ because mother passes on X^h.

We can predict 2 in 4 children will have normal clotting of blood, 1 in 4 will be female and a carrier 1 in 4 will be haemophiliac male.

2

(a) The symptoms of cystic fibrosis can be reduced by gene therapy. Explain this process. [2]

The correct gene for the secretion of fluid from epithelial cells applied by aerosol (in liposomes) through lungs.

(b) Genetic engineering may help reduce the frequency of the cystic fibrosis gene. Which cell would be genetically modified so that **all** nucleated cells of a person would carry the improved gene? [1]

zygote / fertilised egg

Practice examination questions

1 (a) List the criteria which must be satisfied before applying the Hardy–Weinberg Principle. [4]

(b) In a population of 160 small mammals, some had a dark brown coat and the others had a light brown coat. Dark brown (B) is dominant over light brown (b). In the population there were 48 light brown individuals. Using the Hardy–Weinberg equations calculate:

(i) the frequency of homozygous dominant and heterozygous individuals in the population [3]

(ii) how many of each of the genotypes (BB, Bb, bb) would there be in a future population of 10 000 individuals? [2]

[Total: 9]

2 Match each term with its correct definition.

A co-dominance
B polygenic inheritance
C genotype
D polyploid
E somatic

(i) a cell which is not involved in reproduction [1]

(ii) a nucleus which has three or more sets of chromosomes [1]

(iii) a feature which is controlled by two or more genes, along different loci along a chromosome [1]

(iv) two alleles which are equally expressed in the organism [1]

(v) all of the genes found in a nucleus, including both dominant and recessive alleles. [1]

[Total: 5]

3 The letters below represent the organic bases along the coding strand of a DNA molecule.

CCG ATT CGA TAG

(a) What term is given to each group of three bases? [1]

(b) Give **two** functions of a group of three organic bases. [2]

(c) Using the strand of DNA above show **three** different types of point mutation. [3]

[Total: 6]

4 The cell on the right shows a cell at the beginning of telophase II during meiosis.

(a) How many chromosomes were there in the parent cell at the beginning of meiosis? [1]

(b) Describe **one** difference between telophase II and

(i) telophase I of meiosis

(ii) telophase of mitosis. [2]

(c) Describe the stage immediately before telophase II. [2]

[Total: 5]

Chapter 6

Biodiversity

The following topics are covered in this chapter:

- *Classification*
- *Evolution*

- *Manipulation of reproduction*

6.1 Classification

After studying this section you should be able to:

- *indicate sources and effects of variation in organisms*
- *outline features of continuous and discontinuous variation*
- *name the five kingdoms and describe the main features of each group*

LEARNING SUMMARY

What is variation?

AQA A	M5
AQA B	M4
EDEXCEL	M5
OCR	M5
WJEC	M5
NICCEA	M5

Species throughout the biosphere differ from each other.

Variation describes the differences which exist in organisms throughout the biosphere. This variation consists of differences **between** species as well as differences **within** the same species. Each individual is influenced by the environment, so this is another source of variation.

genotype + environment = phenotype

The alleles which are expressed in the phenotype can only perform their function efficiently if they have a supply of suitable substances and have appropriate conditions. Ultimately new genes and alleles have appeared by mutation (see page 81). The spontaneous appearance of **advantageous** new mutations is also possible (see page 81). This may lead to formation of new species.

Continuous and discontinuous variation

Continuous variation

This is shown when there is a range of **small incremental differences** in a feature of organisms in a population. An example of this is height in humans. If the height of each pupil in a school is measured then from the shortest pupil to the tallest, there are very small differences across the distribution. This is shown by the graph below which shows smooth changes in height across a population. This type of variation is shown when features are controlled by **polygenic inheritance**. A number of genes **interact** to produce the expressed feature.

Remember that a species shows continuous variation when there are small incremental differences, e.g. height of people in a town. Beginning with the smallest and ending with the tallest there would probably be at least one person at each height, at 1cm increments. A smooth gradation of differences!

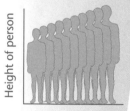

Height of person

Key points from AS

- **Variation**
  Revise AS pages 94–95

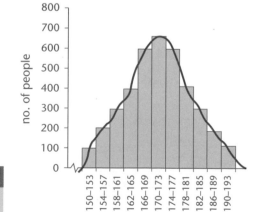

no. of people

Height (in 3cm bands)

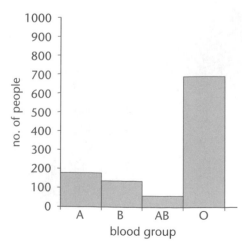

no. of people

blood group

Discontinuous variation

This is shown when a characteristic is expressed in discrete categories. Humans have four discrete blood groups, A, B, AB or O. There are no intermediates, the differences are clear cut!

Classification system

AQA A	M5
AQA B	M4
EDEXCEL	M5
OCR	M5
WJEC	M5
NICCEA	M5

Classification or taxonomy is the way that organisms are divided into groups. The system is based on similarities, differences, and on patterns of evolutionary history (**phylogenetic**). The organisation into groups helps us to identify organisms as we are able to check characteristics against the criteria of a group.

There are five kingdoms

Try this!
Pretty **P**olly **F**inds **P**arrots **A**ttractive.
It will help you to remember the kingdoms!

Most examination boards test knowledge of characteristics across the kingdoms. Information will usually be given for any sub-group tasks.

The table includes some of the main features of each kingdom. Prokaryotae and Protoctista tend to give more of a challenge than some of the other kingdoms, and are examined more often.

Remember that Protoctista, Fungi, Plantae and Animalia are all eukaryotic. They all have membrane-bound organelles, e.g. mitichondria. Flagellae, if present, have a 9 + 2 microtubule structure.

The only similarity with the two examples is that they are in the kingdom Animalia. This is very informative because all Animalia have so common features.

→ increase in complexity

Prokaryotae	*Protoctista*	*Fungi*	*Plantae*	*Animalia*
very simple cells with few organelles	unicellular cells with membrane-bound organelles	heterotrophic nutrition	multicellular organisms which are photosynthetic	multicellular organisms which are heterotrophic
no membrane-bound organelles	some are photo-synthetic, but many have heterotrophic nutrition	some saprotrophic, some parasitic	cells have cellulose cell wall, sap vacuole and chloroplasts	no cell walls no sap vacuoles
if there are flagellae, then not 9 + 2 system of microtubules	reproduction usually involves fission	consists of thread-like hyphae, chitin cell walls	reproduce by seeds, or by spores, some sexual, some asexual.	
DNA in strands, no true nucleus	e.g. algae, and protozoa	many nuclei in hyphae, not in one per cell organisation		
e.g. bacteria and cyanobacteria		reproduction involves the production of spores		

Taxonomy within a kingdom

Each kingdom can be sub-divided into a number of progressively smaller groups. Ultimately this leads to an individual type of organism, a **species**.

The hierarchy of the groups is shown below.

	Example 1	*Example 2*
Kingdom	Animalia	Animalia
Phylum	Chordata	Arthropoda
Class	Mammalia	Insecta
Order	Primates	Lepidoptera
Family	Hominidae	Pieridae
Genus	Homo	Pieris
Species	sapiens	brassica

The seven groups can be difficult to remember. Try the easy way!

KING **P**ENGUINS **C**LIMB **O**VER **F**ROZEN **G**RASSY **S**LOPES

The first letter of each word will help you remember. It is a **mnemonic**. It is a very successful technique! An excellent strategy to aid recall.

What is a species?

If two organisms are able to **breed together**, naturally, and **produce fertile offspring**, then they are from the same species.

KEY POINT

6.2 Evolution

After studying this section you should be able to:

- *understand the process of natural selection*
- *predict population changes in terms of selective pressures*
- *understand a range of isolating mechanisms and how a new species can be formed*
- *understand the difference between allopatric and sympatric speciation*
- *understand adaptive radiation*

Natural selection

AQA A	M5
AQA B	M4
EDEXCEL	M5
OCR	M5
WJEC	M5
NICCEA	M5

Learn this theory carefully then apply it to the scenarios given in your examination. Candidates often identify that some organisms die and others survive, but few go on to predict the inheritance of advantageous genes and the consequence to the species.

Throughout the biosphere communities of organisms interact in a range of ecosystems. Darwin travelled across the world in his ship, the *Beagle*, observing organisms in their habitats. In 1858 Darwin, in association with Wallace, published *On the Origin of Species*. In this book he gave his theory of **natural selection**.

The key features of this theory are that as organisms interact with their environment:

- individual organisms of populations are not identical, and can **vary in both genotypes and phenotypes**
- **some organisms survive** in their environment other organisms **die**, effectively being **deleted from the gene pool**
- surviving organisms **go on to breed** and **pass on their genes** to their offspring
- this **increases the frequency of the advantageous genes** in the population.

Consider these factors

- Adverse conditions in the environment could make a species extinct, but a range of genotypes increases the chances of the species surviving.
- Different genotypes may be suited to a changing environment, say, as a result of global warming.
- A variant of different genotype, previously low in numbers, may thrive in a changed environment and increase in numbers.
- Where organisms are well suited to their environment they have adaptations which give this advantage.
- If other organisms have been selected against, then more resources are available for survivors.
- Breeding usually produces many more offspring than the mere replacement of parents.
- Resources are limited so that competition for food, shelter and breeding areas takes place. Only the fittest survive!

What is selective pressure?

In this example the fact that the numbers of herbivores decrease is *another* selective pressure. This time numbers of predators may decrease.

This is the term given to a factor which has a direct effect on the numbers of individuals in a population of organisms, e.g.

'It is late summer and the days without rainfall have caused the grassland to be parched. There is little food this year.'

Here the **selective pressure** is a **lack of food** for the herbivores. Species which are **best adapted** to this habitat **compete** well for the limited resources and go on to survive. Within a species there is a further application of the selective pressure as weaker organisms perish and the strongest survive.

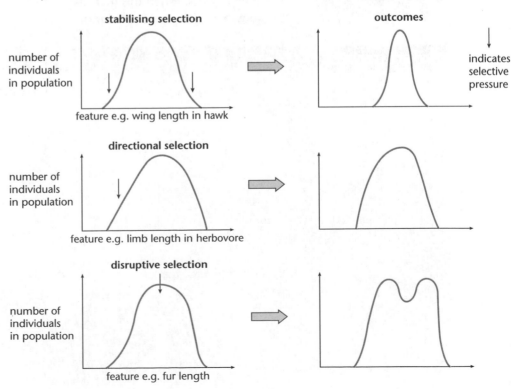

Variation and natural selection

New genes can appear in a species for the first time, due to a form of mutation. Over 1000s of years repeated natural selection takes place, resulting in superb adaptations to the environment.

- The Venus fly trap with its intricate leaf structures captures insects. The insects decompose, supplying minerals to the mineral deficient soil.
- Crown Imperial lilies (*Fritillaria*) produce colourful flowers, and a scent of stinking, decomposing flesh. Flies are attracted and help pollination.
- The bee orchid flower is so like a queen bee that a male will attempt mating.

Considering the above examples, it is no wonder that candidates seem to consider that the organisms actively adapt to develop in these ways. They suggest that the organisms themselves have control to make active changes. **This is not so! There is no control, no active adaptation.**

> ***New genes appear by CHANCE!***   **KEY POINT**

Selective pressures and populations

To find out more about the effects that selective pressures can have, the **normal distribution** must be considered. The distribution below is illustrated with an example.

The mean value is at the peak. There are fewer tall and short individuals in this example. A taller plant intercepts light better than a shorter one.

The further distributions below show effects of selective pressures (shown by blue arrow). Each is illustrated with an example.

Selective pressure at both ends of the distribution cause the extreme genotypes to die. This maintains the distribution around the mean value. Mean wing-length better for flight, better for prey capture.

Selective pressure results in death of less fast animals. May die out due to predators. Faster ones (longer legs) pass on advantageous genes. Distribution moves to right as average individual now faster.

Selective pressure results in death of organisms around mean value. In time this can lead to two distributions. Long fur is adapted to a cold temperature and short fur to a warm temperature. Mean suited to neither extreme.

How can a population become isolated?

The previous example of disruptive selection showed how two extreme genotypes can be selected for. Continued selection against individuals around the former mean genotype, finally results in two discrete distributions. This division into two groups may be followed by, say, advantageous mutations. There is a probability that, in time, the two groups will become incompatible, unable to breed successfully. They have become isolated (genetically). Isolation is a key factor in the **development of new species (speciation)**.

Genetic isolation

> There are several different ways in which genetic isolation occurs, from point mutations to polyploidy.

This can occur with a series of mutations over millions of years. The principle of genetic isolation is described above since two groups, with common ancestors cannot breed together. Polyploidy (see page 80) can also give rise to genetic isolation. Here sets of chromosomes increase, conferring advantage to a new variant. An example of this is *Spartina angelica* (cord grass) with 122 chromosomes per cell. This spontaneously arose from the parent species *Spartina townsendii*, which has 61 chromosomes per cell. *Spartina angelica* is unable to breed with the original parent.

Geographical isolation

> This will help you. Different finches evolved on different islands, but they did have a common ancestor.

This takes place when a population becomes divided as a result of a physical barrier appearing. For example, a land mass may become divided by a natural disaster like an earthquake or a rise in sea level. Geographical isolation followed by mutations can result in the formation of new species. This can be illustrated with the finches of the Galapagos islands. There are many different species in the Galapagos islands, ultimately from a common ancestral species. Clearly new species do form after many years of geographical isolation. This is **allopatric speciation**.

Reproductive isolation

> A new pheromone is produced by several antelopes as a result of a mutation. The mainstream individuals refuse mating as a result of this scent. An isolated few do mate. This is reproductive isolation.

This is a type of genetic isolation. Here the formation of a new species can take place in the same geographical area, e.g. mutation(s) may result in reproductive incompatibility. A new gene producing, say, a hormone, may lead an animal to be rejected from the mainstream group, but breeding may be possible within its own group of variants. When this mechanism results in the production of a new species it is known as **sympatric speciation**.

Progress check

The graph below shows the mean length of roots in *Cirsium arvense*.

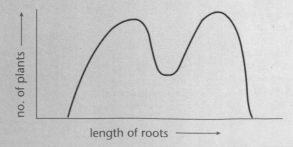

(a) Which type of distribution is shown? Give a reason for your answer.

(b) Name the taxonomic term given to each of the following:

    (i) *Cirsium* (ii) *arvense.*

(b) (i) genus (ii) species

(a) The graph appears to show as disruptive variation, due to two peaks. It also shows as continuous variation since root lengths change in small increments.

How does evolution take place?

Existing species are the result of slow, gradual changes in their ancestors over millions of years. This is **evolution**. Evidence for evolution exists because:

- **fossils** have been found and linked to the modern species
- the **fossil time sequence** is laid down in the **sedimentary layers** beneath the Earth's surface
- **mutations appeared spontaneously** resulting in new advantageous features appearing, often visible in the fossil time sequence
- disadvantageous features often resulted in the disappearance of an organism after a certain period; this is **extinction**
- **vestigial features** in modern species no longer have a function; this points to the fact that ancestors *did* have a use for the feature – **change** (a key idea in the theory of evolution) has taken place
- **common DNA** sequences can be traced across the taxonomic groups and back through to the DNA of well preserved ancestors.

Clearly evolution has taken place in the past and is still taking place!

Genetic conservation

Examiners set questions about genetic conservation. Candidates often give answers which imply that it means the keeping of *genes* in some way, as if they are stored, *detached* from cells.

WRONG! Storage methods include freezing semen or embryos, and cold storage of seeds. Whole cell storage is the current mechanism. Safari parks act as gene pool extensions.

A supply of genetic material to be used in breeding programmes in future years needs to be retained. Geneticists would benefit from an expanding gene pool to improve domesticated animals and crop plants. The ability to retain old varieties of plants in botanic gardens and keeping 'rare breed' farms is vital so that potentially advantageous genes are always available. We must not lose them! Zoos and safari parks can help. Keeping a breeding line of the organisms active is an option but additionally there are less expensive hi-tech methods available. **Sperm**, **embryo**, and **seed banks** can be used for long-term storage.

Deep-freeze storage is not such a recent invention as we at first thought. Glaciers from a previous Ice-age encapsulate a range of carcasses of extinct organisms. Global warming defrosts outer layers and reveals preserved dead organisms. Immediately on exposure to the air the organic material begins to decompose and flies are attracted. Scientists currently search for pre-historic animals such as the mammoth. Retrieving perfectly preserved DNA is the aim. Insertion of mammoth DNA into a zygote of another animal, e.g. an elephant, may allow the re-emergence of this extinct animal.

modern elephant

mammoth

How can we prevent the extinction of organisms?

Wild-life reserves are needed in both terrestrial and aquatic habitats.

The insatiable demand for land to satisfy human requirements is a major problem. Reduced habitat areas means that there is much less land for the organisms to exploit. Modern breeding technology can be employed, e.g. in China the giant panda has problems in that its bamboo habitat is drastically reduced. With numbers very low, and extinction looming, **artificial insemination** has come to the rescue with dramatic success!

6.3 Manipulation of reproduction

After studying this section you should be able to:

- *understand the processes of artificial insemination and embryo transplant*
- *understand the need for progeny testing*
- *understand the problems of inbreeding*

Artificial insemination (AI)

EDEXCEL	M5
OCR	M5
WJEC	M5
NICCEA	M5

AI can be used for a range of different animals. It is even used in humans. Soldiers leaving for action often leave a genetic insurance, their semen!

Important points

- Some domesticated animals have been selectively bred to produce desired features for human needs but which prevent breeding by natural means, e.g. modern turkey breeds. Their breast muscles are so big that the physical act of mating is impossible.
- In many animals a male has limitations as to the number of females he can physically breed with. Males produce millions of sperms, semen dilution allows many more offspring to be produced. Cost effective!
- Top quality, disease-free animals only are used.
- Semen from a variety of males can be used to inseminate a population which avoids inbreeding.
- Even when an animal has died the frozen semen is still viable.

AI in cattle (UK)

Progeny testing

This process allows for the systematic testing of offspring and the comparison of data. Only when a sire (male) has successfully demonstrated that he can transmit suitable features in progeny (offspring) would his semen be used on a wider commercial scale. In the UK a bull is seven years old before his semen becomes available for the national AI programme. His progeny must have performed successfully. Some bulls fail and semen from the seven years would be destroyed. Only genetically superior bulls are used!

Preparation and storage of semen

Another advantage of AI, easy transport!

SEMEN WORLD LTD

EXPORTS WORLD WIDE

- Semen from one ejaculation is diluted around ×50 with milk or egg yolk.
- Glycerol is added as a cryoprotectant, preventing damage by ice formation.
- Sugar is added as an energy source for sperms.
- The external solution has an identical water potential with the sperm contents to prevent osmotic damage.
- Buffer is added to prevent damage due to pH changes.
- Antibiotics are added to destroy spoilage organisms.
- Storage is in straws, kept at −196 °C, in liquid nitrogen.

How are cows artificially inseminated?

- The cow must be in oestrous, having ovulated.
- A straw is thawed out in warm water and inserted into a catheter.
- The catheter is placed in the vagina up to the cervix where the semen is ejected.

A catheter

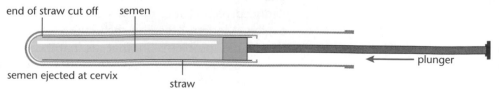

end of straw cut off   semen

semen ejected at cervix

straw

plunger

Superovulation and embryo transfer

These processes can be described with reference to cattle:

- Only genetically superior cows are used in this process.
- Normally a cow produces one ovum at each ovulation.
- Treatment with hormones results in several ova being produced at ovulation.
- Cows would show the signs of oestrous and immediately be artificially inseminated, so that fertilisation takes place.
- This is followed up a week later by the harvesting of embryos.
- This is done with a catheter which does a uterine sweep.
- Embryos are then frozen and kept for later, or artificially implanted into a surrogate cow.
- Transport is very cheap. Formerly a herd of cattle could be exported in a ship at a very expensive cost. Now the same herd is shipped in a liquid nitrogen bucket!
- In this way a Charolais cow can give birth to a pedigree Friesian calf!

Superovulation techniques are used with other animals. These include humans! Some women have fertility problems. Use of hormones to stimulate ovulation yields a number of ova. After *in vitro* fertilisation (in a test tube) resulting embryos can be immediately implanted or frozen for later.

Modern reproduction techniques and genetic engineering will continue to hit the headlines during the lifetime of this book. There are many potential uses and refinements. Ethics and morals must always be balanced against potential benefits, and possible problems.

Micropropagation

New plant varieties are produced by artificial selection. After a long, expensive, breeding programme **just one individual plant** may be produced, e.g. a Day lily (*Hemerocallis*) of new petal colour and disease resistance. **Asexual techniques** must be used to **clone** the variety. In the past techniques such as taking cuttings would have been used and it would take a **number of years** to build up enough plants for a commercial launch. It is now possible to replicate plants by another asexual method, **micropropagation**.

The diagrams below outline the process.

Advantages of micropropagation

- Generating new plants from the apical meristem tissue eliminates many plant viruses, so usually, virus free plants are produced.
- If the material is available the process can take place at any time of the year.
- Even a tiny explant or callus can be cut into pieces and sub-cultured.

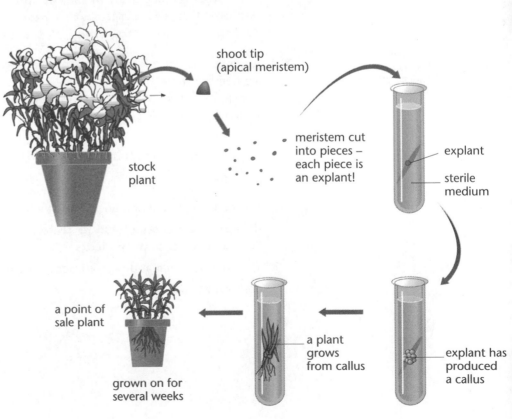

stock plant

shoot tip (apical meristem)

meristem cut into pieces – each piece is an explant!

explant

sterile medium

explant has produced a callus

a plant grows from callus

a point of sale plant

grown on for several weeks

Sample question and model answer

The graphs below show the height of two pure breeding varieties of pea plant, Sutton First and Cava Late.

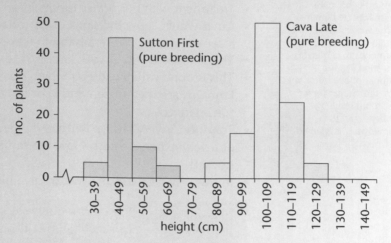

Continuous variation can confuse you sometimes when examiners display the data in categories as histograms. **This is not discontinuous!**

(a) (i) Which types of variation are shown by the pea variety, Sutton First? Give evidence from the bar graph to support your answer. [4]

Continuous variation – this is shown by the increase across the distribution (even though the peas are pure breeding).

Environmental variation – shown by the range of different heights.

(ii) Which type of variation is shown **between** varieties Sutton First and Cava Late? Give evidence from the bar graphs to support your answer. [2]

Discontinuous variation – the two distributions are separate and do not intersect.

(iii) Both Sutton First and Cava Late have compatible pollen for cross-breeding. Suggest why they do **not** cross breed. [1]

As implied by the names, Sutton First flowers before Cava Late, so that flowers are not ready at the same time.

When you are asked to 'suggest' then a range of different plausible answers are usually acceptable.

(b) Plant geneticists considered that many years ago the two varieties of pea had the same ancestor.
   (i) Suggest what, in the ancestor, resulted in the difference in height of the two varieties? [1]
   mutation

   (ii) Suggest what caused this change. [1]
   radiation

(c) (i) Define polygenic inheritance. [1]

The inheritance of a feature controlled by a number of genes (not just a gene at one locus!).

   (ii) Which type of variation is a consequence of polygenic inheritance? [1]
   continuous variation

Practice examination questions

1 The following key distinguishes between the five kingdoms.

	Organisms without membrane bound organelles	A
	Organisms with membrane bound organelles	GOTO 2
2	Organisms have hyphae	B
	Organisms do not have hyphae	GOTO 3
3	Organisms unicellular or colonial	C
	Organisms not unicellular or colonial	GOTO 4
4	Organisms multicellular and have thylakoid membranes in some cells	D
	Organisms multicellular and have no thylakoid membranes in any cells	E

Name kingdoms A, B, C, D and E [5]

2 (a) Explain the difference between allopatric and sympatric speciation. In each instance use an example to illustrate your answer. [6]

(b) How is it possible to find out if two female animals are from the same species? [2]

[Total: 8]

In classification questions always look carefully at the information given. Here *some* of the answers are in the stem of the question.

3 (a) The song-thrush (*Turdus ericetorum*) and mistle-thrush (*Turdus viscivorus*) are in the same family, Turdidae. Large sections of their DNA are common to both species. Complete the table to classify both organisms. [3]

	mistle-thrush	*song-thrush*
Kingdom		
Phylum	Chordata	Chordata
	Aves	
	Passeriformes	Passeriformes
Genus		
Species		
	[1]	[1]

[1]

(b) Assuming that the song-thrush and mistle-thrush evolved from the same ancestry group which type of selection took place to produce the two species? [1]

[Total: 4]

Chapter 7

Ecology and populations

The following topics are covered in this chapter:

- Investigation of ecosystems
- Behaviour

7.1 Investigation of ecosystems

Measurement in an ecosystem

AQA A	M5
AQA B	M5, M6
EDEXCEL	M5
OCR	M5
NICCEA	M4

The study of ecology investigates the inter-relationships between organisms in an area and their environment. The components of an ecosystem can be measured using a variety of techniques.

Estimating populations

Point quadrat

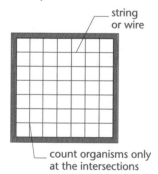

string or wire

count organisms only at the intersections

Often a full count of organisms in an ecosystem is not possible because of the size of the ecosystem. A **sampling technique** is used which requires a **quadrat**. This is a small area enclosed by wire or wood, around 0.25 m². When placed down in the ecosystem the organisms inside the area can be counted, as well as the **abiotic factors** which influence their distribution. Ecologists use units to measure organisms within the quadrats. Frequency (f) is an indication of the presence of an organism in a quadrat area. This gives no measure of numbers. However the usual unit is that of density, the numbers of the organism per unit area. Sometimes percentage cover is used, an indication of how much of the quadrat area is occupied.

Consider a survey of two species *Taraxacum officinale* (dandelion) and *Plantago major* (Great plantain) of the lawn habitat shown below.

A simplified results table

Quadrat no	Dandelion
1	2
2	12
3	15
4	3
5	4
6	8
7	7
8	10
9	9
10	15

mean = 8.5 per quadrat
Dm^{-2} = 34

Lawn habitat

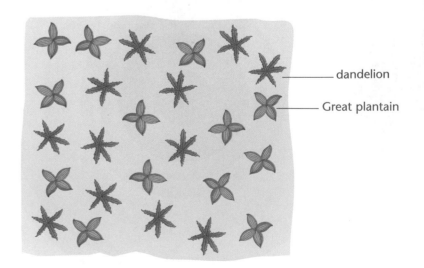

dandelion

Great plantain

Key points from AS

- **What is an ecosystem?**
  Revise AS page 110

Optimum number of quadrats

The rule which dictates how many quadrats should be used is the point when additional quadrats do not significantly change the average or mean value.

It is important to use a suitable technique when surveying with quadrats. When you observe a habitat which appears **homogeneous** or **uniform**, like a field yellow with buttercups, then you should use **random quadrat placement**. The area should be gridded, numbers given to each sector of the grid, then a random number generator used. The probability of the numbers in one quadrat representing a field would be very low. In practice the **mean** numbers from large numbers of quadrats do represent the true numbers in a habitat.

Belt transects

This term is given to another quadrat technique. This method should be used when there is a **transition** across an area, e.g. across a pond or from high to low tide on the sea shore. Use belt transects where there is **change**. The belt transect is a line of quadrats. In each quadrat a measurement such as density can be made. One transect is not enough! Always do a number of transects then find an average for quadrats in a similar zone.

A simplified results table

Quadrat no	flag iris	water lily
1	10	0
2	7	0
3	1	0
4	0	5
5	0	4
6	0	0
7	0	5
8	0	3
9	0	0
10	1	0
11	8	0
12	4	0

This is just one belt transect. A number would be used and an average taken for each corresponding quadrat.

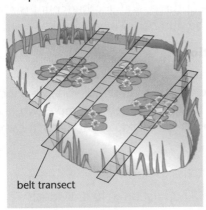

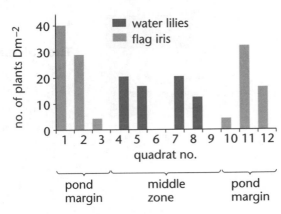

A bar-graph would be used to show the **distribution** of plant species across the pond. Note that there would be more than just two species! The graphs show how you could illustrate the data. Clearly flag irises occupy a different niche to water lilies.

Other uses of quadrats

They can also be used to survey animal populations. It is made easier if the organisms are **sessile** (*they do not move from place to place*), e.g. barnacles on a rock. In a pond the belt transect could be coupled with a kick sampling technique. Here rocks may be disturbed and escaping animals noted. Adding a further technique can help, such as using a catch net in the quadrat positions. The principle here is that the techniques are **quantitative**.

KEY POINT

Measuring factors

Graphical data can show relative numbers and distribution of organisms in a habitat. The ecologist is interested in all factors, **biotic and abiotic**, which influence those organisms. Listed below are some of the factors which may be measured:

- carbon dioxide level
- oxygen level
- pH
- light intensity
- mineral ion concentration
- level of organic material.

There are many more factors. These are just a selection.

Datalogging

Modern ecology is so much more convenient and accurate than conventional techniques! Datalogging relies on environmental probes, interfaces and computers to measure and display data. Measuring factors over a complete day leaves no data-gaps! Fluctuations in light reaching a plant can be monitored over 24 hours and related to photosynthetic patterns. A light probe would be employed, but there are many more.

Capture, mark, release, recapture

This is a method which is used to estimate animal populations. It is an appropriate method for motile animals such as shrews or woodlice. The ecologist must always ensure minimum disturbance of the organism if results are to be truly representative and that the population will behave as normal.

The technique

- Organisms are captured, *unharmed*, using a quantitative technique.
- They are counted then discretely marked in some way, e.g. a shrew can be tagged, a woodlouse can be painted (*with non-toxic paint*).
- They are released.
- They are recaptured, and another count is made.
- This gives the number of marked animals and the number unmarked.

The calculation

S = total number of individuals in the total population.
S_1 = number captured in sample one, marked and released, e.g. 8.
S_2 = total number captured in sample two, e.g. 10.
S_3 = total marked individuals captured in sample two, e.g. 2.

$$\frac{S}{S_1} = \frac{S_2}{S_3} \qquad \text{so, } S = \frac{S_1 \times S_2}{S_3}$$

$$S = \frac{8 \times 10}{2} \quad \text{population} = 40 \text{ individuals}$$

Remember the equation carefully. You will **not** be supplied with it in the examination, but you will be given data.

The index of diversity

This is used as a measure of the range and numbers of species in an area.

$$\text{index } d = \frac{N(N-1)}{\Sigma\, n(n-1)}$$

N = total no. of all individuals of all species in the area
n = total no. of individuals of one species in an area
Σ = the sum of

Consider this example of animals in a small pond

crested newt	8
stickleback	20
leech	15
great pond snail	20
dragonfly larva	2
stonefly larva	10
water boatman	6
caddisfly larva	30
	N = 111

$$d = \frac{111 \times 110}{(8\times7)+(20\times19)+(15\times14)+(20\times19)+(2\times1)+(10\times9)+(6\times5)+(30\times29)}$$

$$d = \frac{12\,210}{2018} \quad \text{so } d = 6.05$$

Before using the technique you must be assured that:
- there is no significant migration
- there are no significant births or deaths
- marking does not have an adverse effect, e.g. the marking paint should not allow predators to see prey more easily (or *vice versa*)
- organisms integrate back into population after capture.

Remember that the method is suitable for large population size only.

In another pond there were:

crested newt	45
stickleback	4
leech	18
great pond snail	10
d = 2.6	

Look at both indices. 6.05 is an indicator of greater diversity. The higher number indicates greater diversity.

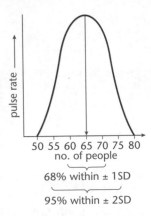

Standard deviation

Standard deviation (SD)

This is a measure of the **spread** of results at **either side of the mean** in an investigation. Consider this example: the pulse is taken from each person at rest, in a group of people. After collecting the results the mean or average is calculated. The mean does not give an indication of the spread of results. For this, standard deviation is needed.

$$\text{Standard deviation} = \sqrt{\frac{\Sigma d^2}{n}}$$

Σ = the sum of

d = difference between each value (e.g. individual pulse) and mean (pulse)

n = total number of readings (e.g. the number of pulses taken)

Statistically it has been shown that 68% of all readings are within + or − one standard deviation and that 95% of all readings are within + or − two standard deviations. Small SDs show that most readings are within a narrow range. Large SDs show that the mainstream of readings have a greater range.

Ecological conservation

In a world where human population increase is responsible for the destruction of so many habitats it is necessary to retain as many habitats as possible. Ecological surveys report to governments and difficult decisions are made. Fragile habitats like the bamboo woodlands of China support a variety of wildlife. Conservation areas need to be kept and maintained to prevent extinction of organisms at risk. In the UK we have **sites of special scientific interest** which are given government protection.

Conservation requires management

Although the word itself implies to 'keep' something as it is, much effort is needed. An area of climax vegetation, e.g. oak woodland, is less of a problem, since it will not change if merely left to its own devices. However, many of the seral stages, e.g. birch woodland along the route to climax, require much maintenance.

> In an examination you may be given data to analyse. Always consider the plant life which is needed to support herbivores, as well as predators further along the food chains.

Animal populations need our help, especially when it is often by our own introduction that specific species have colonised an area. Deer introduced into a forest may thrive initially but due to an efficient reproductive rate exceed the carrying capacity of the habitat. **Carrying capacity** is the population of the species which can be adequately supported by the area.

> What effect would an increase in mesh size of trawler nets have on the fish catch?

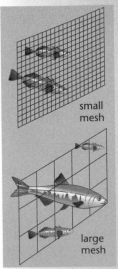

small mesh

large mesh

The bigger the mesh the more fish escape.

Sometimes herbivores could cause destruction of their habitat by overgrazing, and so must be **culled**. **Predators** could be introduced to reduce numbers, but they also may need culling at some stage. **Difficult decisions** need to be taken. In the aquatic habitats similar problems exist. Cod in the North Sea is being harmed by over-fishing. Agreements have been made by the EEC to **reduce fishing quotas** and create **exclusion zones** to **allow fish stocks to recover**. Even before this agreement smaller fish had to be returned to the sea after being caught to increase the chances of them growing to maturity and breeding successfully.

Endangered species require protection

All over the world many animals and plants are at the limits of their survival. The World Wide Fund for Nature is a charity organisation which helps. The organisation receives support from the public and artists such as David Shepherd. He gives donations from the sale of all of his wildlife paintings, helping to maintain the profile of animals so that we invest in survival projects like protected reserves.

Interspecific competition takes place when **different** species share the same resources.

Intraspecific competition takes place when the **same** species share the same resources.

7.2 Behaviour

After studying this section you should be able to:

- *describe innate behaviour, kinesis, and taxis*
- *understand habituation and imprinting*
- *describe a range of territorial behaviour*

The behaviour of organisms

AQA A ▶ M6
AQA B ▶ M6
OCR ▶ M5

Organisms respond to the **biotic and abiotic factors** of their environment. Biotic factors include response to other species, e.g. the feeding behaviour of grouse from heather on moorlands and the use of the heather to hide from predators. The grouse also respond to each other, e.g. in courtship display. There are different types of behaviour.

Innate behaviour

This behaviour is '**pre-programmed**' by an organism's **genes**. When analysing behaviour it is difficult to determine whether it is innate or learned.

It is safe to say that immediately after the birth of a baby the 'sucking' action to obtain milk from mother's mammary glands is innate. Similarly, the pecking behaviour of a chicken, whilst still in an egg, to break the shell, must be innate. As an animal gets older it may well develop patterns of behaviour learned from its experiences. It becomes more and more **difficult to categorise** the behaviour.

Kinesis

This takes place when the response of an organism is **proportional to the intensity of a stimulus**. Kinesis takes the form of an **increase in movement**, but this is **non-directional**. An example of kinesis is shown by woodlice. Intense heat which would harm the woodlice causes them to increase speed and move in random directions. In this way some of the population have a **greater chance of survival**.

A second example is shown by woodlice. They respond to a dry environment by increasing random movements but slow down if they reach high humidity.

Taxis

This is a **directional response to a stimulus**. It can be a **positive taxis**, towards, or **negative taxis**, away. An example can be seen using a microscope to observe a group of living specimens of *Euglena viridis*. This is a protoctistan which photosynthesises. Individuals swim to an air bubble and cluster around to obtain maximum CO_2 for photosynthesis. This is **positive chemotaxis** because the organism moves towards the CO_2 source.

Progress check

(a) List the abiotic factors you may need to measure in a pond survey.

(b) How could you take measurements most efficiently, over a 24-hour period.

(b) use of environmetal probes, interface and computer
(a) oxygen, carbon dioxide, pH, light, temperature, mineral ions

Learning

This takes place when an organism changes behaviour as a result of experience within the environment. As a result of the experience future behaviour becomes modified. For example, a pupil misbehaves and is placed on detention. The pupil learns (hopefully!) that the behaviour should not be repeated. The detention is negative reinforcement. Perhaps positive reinforcement is better to support good behaviour!

Conditioned reflexes

Pavlov experimented with dogs.

- He checked that the group of dogs did not produce saliva when he rang a bell at a time not related to feeding. (**Control**)
- He fed groups of dogs at a specific time each day.
- He measured the amount of saliva produced just before they were fed.
- He then began to ring a bell just before giving the food.
- The dogs began to salivate profusely.
- The bell would elicit exactly the same response as the original stimulus.
- After a while the level of salivation decreased if the food reward was not given.
- Without **positive reinforcement** the level of response would finally disappear completely.

We are conditioned to respond to advertising in a similar way. A cola drink advertisement uses the latest 'rock' song and glamorous models. We go to the supermarket and respond by buying the product, relating it to the pleasurable experience of the advertisement. Repeat purchases will only continue if the taste of the cola elicits a positive taste perception.

Habituation

This takes place when an organism is subjected to a **stimulus which is not harmful or rewarding**. As a result of continued subjection to a stimulus a **response will gradually decrease** and can finally disappear completely. A farmer puts an electronic bird scarer into a field. Birds are frightened off by frequent 'bangs'. They return, gradually more closely and finally have learned that the scarer is non-threatening. Soon they feed close to the scarer which has no effect. This is **habituation**.

Advertisements have a short 'shelf-life'. Continued exposure to the same advertisement results in habituation so that the response decreases. No wonder media advertising is replaced every few weeks!

The Sand Hill Crane and imprinting

This endangered species is reared in incubators and re-introduced into the wild. There is a problem! Young cranes would imprint upon humans, so when re-introduced would move towards people. Dangerous! Each day keepers dress up in 'crane' uniforms. In the wild the birds then move towards groups of adult cranes.

Examples are given to illustrate **principles**. It is unlikely that you will be given the same examples in your examination. Apply the principles to the given data.

Imprinting

This takes place during the very early life of an organism, e.g. a chick emerges from its egg shell and immediately **bonds** with a close-by object. In nature, this will be the mother hen. The mother hen will impart useful behavioural patterns to the youngster, thus having **survival value**. From an incubator the focus of the imprinting would be a human. The imprinting behaviour is that the chick, in this instance, will follow the human or any object to which it is first exposed.

Territorial behaviour

Populations of organisms living in an area can benefit from territorial behaviour. Too many animals of the **same species**, living in an area, **competing** for food would put the whole population in danger. Many species display territorial behaviour which prevents this outcome.

What is the advantage of male aggression to other males in a population?

The fittest organisms need to pass on their advantageous genes to offspring. In deer herds a dominant stag (male) is challenged by a younger male occasionally. Antler to antler fights take place and there is potential damage to both. In time a new dominant stag takes over the family group and now has exclusive mating rights with a group of hinds (females). This behaviour ensures the male reproductive role involves only the strongest males. The gene pool is improved!

Defence of the territory

- Animals are often **aggressive to members of the same species**, outside of the same family group.
- Territory is demarcated in a variety of ways, such as marking with **urine**, **faeces**, or **scent**. Birds use **song**, whereas other animals have characteristic **calls**. Excluding others in these ways **can prevent physical confrontation** which often results in injury.
- It is an advantage to the species to have a **feeding range** which excludes others. This increases the chances of there being **enough food for the family group**.
- The apportioning of territories serves as **density dependent regulation** so that the best use is made of existing resources.
- A further advantage is that a territory marks out a designated **mating area**. Other males will usually remain outside of the zone. Offspring have protection for their early development.

Courtship behaviour

This behaviour is species dependent and courtship display is anchored in the genes (innate behaviour see page 114). Courtship rituals are very important to ensure that:

- the opposite sexes recognise each other
- the animals will mate with organisms from the same species (mating is more likely to produce fertile offspring)
- the act of mating is synchronised with the oestrous cycle. In pigs a boar is always ready to mate but a sow is only receptive to him at ovulation. She produces pheromone attractants to encourage the boar.

Sign–stimulus release factors

These are environmental cues which trigger patterns of innate behaviour in organisms.

Example 1 – the stickleback

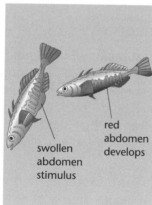

swollen abdomen stimulus

red abdomen develops

In springtime as the mean temperature and day length increase, a **colour change** in stickleback males takes place. The ventral surface becomes red and they go on to build nests and defend their territory. Clearly the environmental cues stimulate physiological and behavioural changes. A **zig-zag ritual** by the male elicits a female to display her swollen abdomen to the male. The process culminates in fertilisation.

Example 2 – delayed implantation

Animals such as badgers mate at a specific time of year. Environmental cues bring females into season, November in this instance. After fertilisation, implantation into the endometrium is synchronised in all the females regardless of the day of mating. The sychronous timing of implantation ensures that the offspring are born when there is a spring flush of food. Survival is more likely!

Environmental cues trigger migratory behaviour, e.g. shortening day length elicits preparatory behaviour of swallows, which collectively fly off to South Africa. Similarly, cues stimulate the return journey.

Sample question and model answer

When given a passage, line numbers are often referred to. Try to understand the words in context. Do not rush in with a pre-conceived idea!

Read the passage, then answer the questions below.

line 1   Around the UK coast there are two species of barnacle, *Chthamalus stellatus* and *Balanus balanoides*. Both species are sessile, living on rocky sea shores.

The adult barnacles do not move from place to place but do reproduce *line 5*   sexually. They use external fertilisation. Larvae resemble tiny crabs and are able to swim. At a later stage these larvae come to rest on a rock where they become fixed for the remainder of their lives.

The barnacles are only able to feed whilst submerged.

Adult *Chthamalus* are found higher on the rocks than *Balanus* in the adult *line 10*   form as shown in the diagram below. Scientists have shown that the larvae of each species are found at all levels.

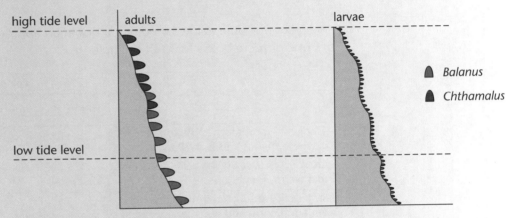

(a) Name the genus for each barnacle.   [1]

    Chthamalus and Balanus

In this question are key terms, of which you will need to recall the meaning. This is only possible with effective revision. Did you already know the key terms **genus**, **sessile** and **motile**? Knowledge of these terms would enable access to other marks, only easy when you have key word understanding. Try writing out a glossary of terms to help your long-term memory.

(b) What does the term sessile mean? (line 2)   [1]

    is not motile, i.e. does not move from place to place

(c) Suggest how it is possible for neighbouring *Balanus* individuals to breed sexually (line 3) with each other even though they are sessile.   [1]

    produce sperms which swim through the water

(d) Explain **one** advantage of the larvae being motile.   [2]

    able to colonise new areas

    where there may be more nutrients

(e) Which type of competition exists between *Chthamalus* and *Balanus*?   [1]

    interspecific competition

(f) Suggest an explanation for the distribution of each species of barnacle.   [6]

    The larvae are found at all tide levels

    at a lower tide level the barnacles are submerged for longer

    Balanus may grow at a faster rate can compete for food better than Chthamalus, which dies out at lower levels

    near higher tide level the barnacles are exposed to open air for longer Balanus may not be adapted to withstand desiccation, whereas Chthamalus can withstand drying out, and so survives without the competition from Balanus.

Practice examination questions

1 Ecologists wished to estimate the population of a species of small mammal in a nature reserve.

- They placed humane traps throughout the reserve and made their first trapping on day one, capturing 16 shrews.
- They were tagged then released.
- After day four a second trapping was carried out, capturing 12 shrews.
- Five of these shrews were seen to be tagged.

(a) The ecologists must be satisfied of a number of factors before using the 'capture, mark, release, recapture' method. List three of these factors. [3]

(b) Use the data to estimate the shrew population. Show your working. [2]

(c) Comment on the *level* of reliability of your answer. [1]

[Total: 6]

2 Complete the table below by putting a tick in an appropriate box. You may tick one or more boxes for each example.

	Type of behaviour			
	kinesis	*innate*	*positive taxis*	*negative taxis*
A bolus of food reaches the top of our oesophagus and is swallowed.				
An insect moves from a cold, dry area to a warm, humid one.				
Springtail are subjected to increasingly hot conditions, and react by increasing speed in a number of directions. Some go towards the heat source and die.				
A queen bee accepts the advances of a drone bee and is mated.				
A motile alga swims towards light.				

[5]

3 A Grebe is a water bird which displays a distinctive courtship ritual. Male behaviour is distinctive from that of the female.

State **three** advantages to the species of this behaviour. [3]

Chapter 8

Further effects of pollution

The following topics are covered in this chapter:

- *Water pollution*
- *Air pollution*
- *Pollution control*

8.1 Water pollution

After studying this section you should be able to:

- describe the sources and effects of a range of aquatic pollutants
- understand how the presence of indicator species gives signs of the degree of pollution
- determine biochemical oxygen demand (BOD)
- understand how water is treated prior to domestic supply
- describe a range of methods to reduce, prevent or avoid aquatic pollution

LEARNING SUMMARY

Sources of water pollution

AQA A	M5
AQA B	M6
EDEXCEL	M5
OCR	M6
NICCEA	M4

A number of domestic, agricultural and industrial processes contribute to water pollution. The following sources result in pollutants reaching habitats such as rivers, and ultimately the sea.

KEY POINT

- **Heavy metal** ions from industrial outflows.
- **Oil spillage** from tanker accidents in the sea.
- **Suspended solids** (e.g. china clay from pottery industry outflows).
- **Excess mineral ions** such as **phosphates** and **nitrates** in sewage, domestic **drainage effluent**, and **fertilisers**.
- **Pesticides**, including **organophosphates**, and **herbicides**, via run-off and leaching.
- **Thermal pollution** as coolant water is returned to rivers from power stations.

industrial outflow   legislation is important!

regular consumption of contaminated food accumulates toxins

Heavy metal ions

These can enter water accidentally or by discharge of industrial waste. In Minimata, Japan, a mercury compound was discharged into the sea-water. It was absorbed by shell-fish which were not, themselves, killed. However, many shell-fish were consumed by people. The toxic mercury ions were passed through the food chains. Since **many** shell-fish were consumed the **toxin built up**. This is **bio-accumulation**. The human consumers suffered a number of severe problems with their **nervous systems**.

Oil spillage

This takes place as a result of accidents. Sea-going tankers, full of crude oil, occasionally have accidents and spill their oil cargo into the sea. This is potentially catastrophic as **oil floats**. Oxygen diffusion from the air is inhibited by the resultant

Key points from AS

- **Effects of human activities on the environment**
  Revise AS pages 119–121

oil

huge numbers of bacteria

fish dead

bottom dwelling plants are rotting

oil slick. The aquatic organisms often die due to lack of oxygen and plants lack light for photosynthesis. Beneath the oil **putrefaction** can take place. **Bacterial action** leads to the highly toxic gas, **hydrogen sulphide** with its rotten egg smell.

Additionally oil affects the feathers of water birds. Contamination can result in their inability to fly and they lose their buoyancy. Complete food chains are destroyed!

particles block the internal siphons of the mussel

Suspended solids

Whenever there are large amounts of particles in water there are problems for many organisms. Suspended particles intercept light, reducing photosynthetic productivity which has a consequential, adverse effect on consumers in food chains.

The pottery industry has contributed to this problem in the past. Fine particles such as waste china clay are particularly harmful. The particles are so small that they are taken in by bivalves such as mussels. These filter feeders cannot cope with the sheer volume of particles and perish as their feeding and gaseous exchange mechanisms are inhibited.

If particles continue to silt up the bottom of the water habitat then many more organisms are harmed by the continued coverage, e.g. bloodworms (Tubifex).

Excess mineral ions

Low amounts of **nitrates** and **phosphates** would be beneficial to aquatic plants. **In excess** they are harmful and cause **algal blooms** which ultimately result in **eutrophication**. **Phosphates** are in detergents which reach rivers via domestic drainage water. Additionally, together with **nitrates**, they are in fertilisers which reach rivers via leaching and run-off from fields. In most instances, only 50% of fertiliser is actually taken up by plants. Sewage effluent is another contributory factor in eutrophication.

ALERT! Do not confuse herbicides with pesticides. Herbicides are used to destroy weeds and may run-off into rivers. Contact herbicides such as sodium chlorate are highly toxic in the environment. Similarly selective herbicides are used. These kill broad-leaved plants but grasses and cereals are unharmed. They cause uncontrolled growth which results, finally, in death. **Dioxins** are very toxic indeed and are a contaminant of the herbicide production. Entering aquatic food chains is a considerable danger!

Pesticides

These include **insecticides**, **fungicides**, and **molluscicides** (used against snails and slugs). It is important that they are **specific** and kill only the pests. Unfortunately they tend to affect organisms other than the target pests. Run-off and leaching again take them to the rivers. **Persistent** pesticides such as **DDT** (an organochlorine) remain in the ecosystem for long periods. Insects affected by DDT enter food chains. The persistent DDT accumulates as it builds up in fatty tissue. DDT passes through rivers and seas reaching all over the world, even in places where it is not used. Banned in the UK many years ago it still remains in the soil and can be detected in animal tissues. In some countries it is still used!

Accumulation along food chain

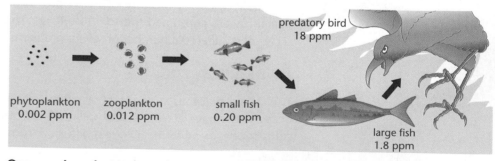

predatory bird 18 ppm

phytoplankton 0.002 ppm

zooplankton 0.012 ppm

small fish 0.20 ppm

large fish 1.8 ppm

Most medicated shampoos to destroy head lice, contain organophosphates! It is dangerous to use them regularly.

Organophosphates have been used to replace the **organochlorines**, being less persistent. Unfortunately they are toxic to humans as well as insects. In humans they inhibit the action of **acetylcholine esterase** at synapses. Used as sheep dip insecticide they are very dangerous if they contact humans and other organisms.

Typically if these dangerous chemicals reach water habitats they may be accumulated along food chains quickly.

Biochemical oxygen demand (BOD)

OCR   M5

This is a quality control method and takes place as follows:

- samples of water or effluent are collected
- each sample is incubated at a temperature of 20 °C for five days
- the amount of dissolved oxygen in mg litre^{-1} or g m^{-3} is the BOD value
- the more microorganisms in the sample the greater is the BOD and the depletion of oxygen is correspondingly high
- the greater the BOD of the effluent the greater is its ability to reduce the dissolved oxygen in a sample.

If investigating BOD of river water, remember:
(a) take samples from different positions – BOD may be variable according to depth
(b) always take an average from each position to ensure that your results are truly representative.

High BOD indicates a low amount of oxygen in the sample!

BOD examples

- Silage effluent seeping into a river = 45 000 g m^{-3}
- Domestic sewage = 348 g m^{-3}
- Non-polluted river water = 2.5 g m^{-3}
- No wonder the fish die when there is a lot of organic pollution around!
- No wonder farmers are fined heavily if sewage effluent seeps into water!
- Legislation helps!

Water treatment

Before being suitable for consumption water needs to be treated. This takes place as follows:

Screening Grids filter out debris such as dislodged weeds.

Sedimentation Water is pumped to a sedimentation tank. Here any heavy particles sink to the bottom. From here the sediment is pumped away.

Flocculation Ferrous sulphate is added as a flocculant which allows many of the remaining organic particles and bacteria to stick together.

Ultra-violet light The water is now exposed to ultra-violet light which kills many microorganisms.

Filtration Water is then filtered through layers of graded sand particles (fine particles at the top and more coarse towards the base). Water percolates through this filter bed slowly. Organic debris is held back in the upper layers so that:

- algae in the uppermost layer take up NO$_3^-$ and PO$_3^-$ released by saprobionts.

Under the algae at lower levels are saprobionts which decompose the organic material releasing the NO$_3^-$ and PO$_3^-$ ions.

The algal layer grows and is occasionally removed.

algae
saprobionts
water
ferrous sulphate
purified water out
flocculated particles sink (removed later)

Note that responses to questions at A Level are often expected to be more complex, e.g. consequential effects are needed, not just 'kills filter feeder'.

Sterilisation

During the previous stage the microbial populations increase. During this stage **chlorine** or **ozone** is added to kill many of the **microorganisms**. The process does not destroy all microorganisms and is best described as **partial sterilisation**.

Progress check

The outfall from a pottery works poured suspended particles into a river. Suggest **two** polluting effects.

- Less light reaches aquatic plants, photosynthesis less, less food for consumers along food chains.
- Filter feeders harmed by the volume of particles, which clog up their feeding and gaseous exchange mechanisms.

8.2 Air pollution

After studying this section you should be able to:

- describe sources and effects of a range of air pollutants
- describe the beneficial effects of recycling

Pollution of the atmosphere

AQA A	M5
AQA B	M6
EDEXCEL	M5
OCR	M6
NICCEA	M4

Gases reach the air in a number of ways including combustion of fossil fuels, industrial emissions and microbial processes. Gases reaching the air include sulphur dioxide, carbon dioxide, carbon monoxide, nitrogen oxides, chlorofluorocarbons (CFCs), methane and smoke particles.

Key points from AS

- Effects of human activities on the environment
  Revise AS pages 119–121

Polluting effects

Large amounts of the above gases can have very serious effects. The greenhouse effect is caused by a number of gases but water vapour and carbon dioxide have the greatest effect due to their high volume in the troposphere. CFCs have a stronger potential contribution to the greenhouse effect but are found in smaller quantities.

Industrial emissions

Industries have traditionally been very defensive about admitting to the production of harmful waste gases. Good examples of known dangerous emissions are **polychlorinated biphenyls (PCBs)**. These are used in plastics manufacture. Limited release of these gases occur during manufacture but much more is given off during incineration of plastics waste.

These persistent chemicals build up in animal tissues and have been found in fat cells in seals. PCBs are considered to be carcinogenic.

What about the hole in the ozone layer?

ALERT!

The 'hole in the ozone layer' is completely different to the 'greenhouse effect!' Students are often confused. The reason for this is that ozone and CFCs are both greenhouse gases.

When answering a question about either topic clarify which topic is being examined. There are no marks if you are wrong!

High up in the stratosphere at around 20–30 km is a layer of ozone (O_3). It is very important because when radiation from the sun passes through this layer much of the UV radiation is absorbed.

What effect does UV radiation have on people?

If skin is exposed to UV radiation then skin cancer, cataracts and problems with the immune system can result.

Which pollutants damage the ozone layer?

Chlorine, nitrogen oxides and CFCs are all responsible. Each reacts to break down the ozone, by a variety of mechanisms. Oxygen is a product of these reactions.

Progress check

Outline the processes by which water is purified before being supplied to the public.

Screening to filter out debris. Sedimentation – heavy particles sink to the bottom.
Flocculation – ferrous sulphate is added, particles stick together.
Ultra-violet light – kills many microorganisms.
Filter – water moves through sand particles slowly.
Organic material is decomposed by saprobionts.
Algae use NO_3^- and PO_3^- released by decay.
Chlorination to kill microorganisms.

8.3 Pollution control

After studying this section you should be able to:

* describe methods of controlling pollution
* understand the benefits of recycling and use of renewable energy resources
* understand the importance of legislation and personal responsibility

What can we do about the pollution problem?

AQA A	M5
AQA B	M6
EDEXCEL	M5
OCR	M5, M6
NICCEA	M4

It is easy to state that many potentially harmful chemicals should not be allowed to enter the environment. However, the human population is increasing at a phenomenal rate and resources are needed to support this growth including housing, food and other products. We need to keep pollution under control. Lifestyle and personal responsibility are key factors to consider if we are to control pollution.

* Legislation and international agreement to limit or ban toxic emissions, e.g. SO_2 reduction.
* Car sharing, using low sulphur fuel, use of catalytic converters.
* Buying organic products rather than those which have been sprayed with pesticides.
* Use of biological control which targets specific pests, rather than using pesticides.
* Breeding programmes to reintroduce endangered species in some areas.
* Set up sites of special scientific interest so that the fragile habitats almost destroyed by previous pollution can be conserved.
* Recycling of products such as aluminium cans requires much less energy input than the processing of bauxite ore, and less of the environment is spoiled at mines.
* Composting of organic waste reduces the use of inorganic fertilisers. Slow release of ions and better holding capacity ensures that less ions reach the rivers by run-off and leaching.
* Use of renewable energy sources including solar power, wind power, wave power, tidal power, hydro-electric power. Gasohol, a derivative of photo-synthetic products, produces H_2O and CO_2 as waste. Methane gas produced from decomposing organic waste is an excellent alternative to fossil fuels.
* 'Scrubbing' of waste gases at coal-fired power stations reduces SO_2 emission. Emission gases are mixed with alkaline fluids.

$$CaCO_3 + SO_2 \rightarrow CaSO_3 + CO_2$$

Calcium sulphite produced in the process is dumped. CO_2 is a lesser problem! There could be an additional charge. Perhaps we should be prepared to pay more to safeguard our environment.

Lethal dose 50 (LD₅₀)

These are toxicity tests for chemicals such as insecticides used in the environment. LD_{50} is the amount of substance, given orally, which kills 50% of a population in the laboratory.

The LD_{50} value for DDT in male mice is $500\,mg\,kg^{-1}$ and for females is $550\,mg\,kg^{-1}$. This proves that scientists showed some responsibility. DDT is now a banned persistent *insecticide*. It was not used on mice, but we need data on toxicity to organisms other than the target because they encounter chemicals in their environment.

Some people have ethical objections to LD_{50} testing. Alternative tests using laboratory cultured tissues are also used.

Sample question and model answer

Do not become confused with multiple graph lines! Follow each along and try to think why each one changes. A fall in one line, followed by a rise in another may suggest that a substance is being used up or converted into another!

The graphs below show the levels of some key substances and the relative numbers of organisms.

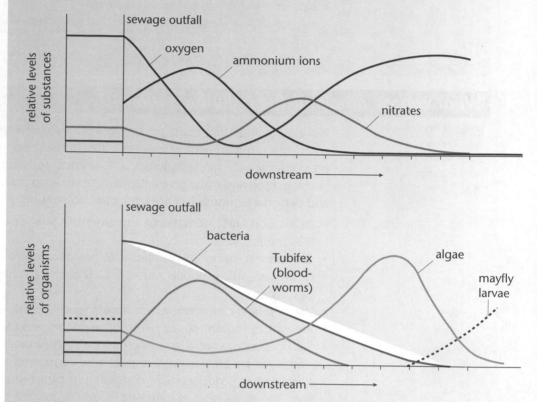

Use the graphs above and your own knowledge to answer the questions below.

(a) Explain the increase of nitrate ions after the sewage outfall. [4]

Sewage is decomposed by saprobiotic bacteria and saprobiotic fungi, this produces ammonium ions.
These are used by nitrosomonas bacteria to produce nitrite (not measured).
Nitrite converted to nitrate by nitrobacter bacteria.

(b) Account for the increase in Tubifex (bloodworm) numbers after the sewage entered the river even though oxygen levels decreased. [3]

Tubifex worms contain haemoglobin.
They absorb oxygen even when it is at low concentration.
Other organisms which normally feed on the Tubifex cannot live with the pollutants.

(c) Which organisms utilise the nitrates? Give reasons for your answer. [3]

Algae.
They need the nitrates to produce proteins.
Peak of algae follows the decrease of nitrates.

(d) Why did the oxygen decrease after the sewage outfall? [1]

Used in aerobic respiration by the saprobiotic bacteria.

Indicator species show the pollution level of an area because of the presence or absence of key species. Mayfly larvae survive in pollution-free water whereas bloodworms in large numbers indicate high sewage levels.

(e) Name **three** indicator species and in each instance suggest an environmental factor associated with their numbers.

Mayfly larvae only present where oxygen is high = low pollution.
Algae are in large numbers where there are nitrates.
Bacteria are in large numbers where organic material (sewage) is high.

Practice examination questions

1 The swan mussel (*Anodonta cygnia*) is a filter feeder. River water is taken into their mantle cavities and past gills before being ejected. Organic particles are moved to the mouth by cilia.

Students investigated the efficiency of particle removal by swan mussels. A colorimeter was used to measure the amount of light absorbed by the water at different stages.

Four tanks were used:

    Tank A – 10 litres non-polluted river water
    Tank B – 10 litres non-polluted river water + 20 swan mussels
    Tank C – 10 litres river water polluted with china clay particles
    Tank D – 10 litres river water polluted with china clay particles + 20 swan mussels

The graph below shows the mean rate of particle removal from water at a temperature of 20°C.

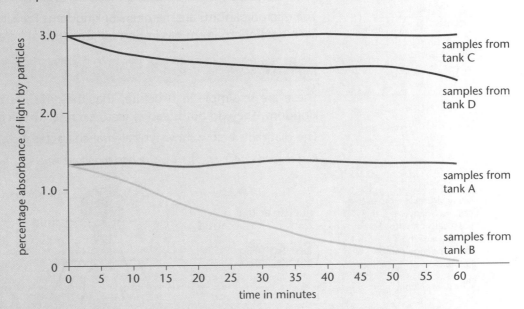

(a) Each tank had a magnetic stirrer. What is the advantage of using a stirrer? [1]

(b) Explain how the units of percentage absorbance can be used to measure the efficiency of filter feeding. [2]

(c) Explain the effect of china clay particles on the efficiency of filter feeding by the mussels. [3]

(d) Suggest **one** effect the swan mussels may have on the plant life of a river. [1]

[Total: 7]

2 Sewage effluent entered a river at two points, A and B, causing eutrophication.
    BOD of sewage effluent at A = $350\,\mathrm{g\,m^{-3}}$
    BOD of sewage effluent at B = $389\,\mathrm{g\,m^{-3}}$

(a) (i) What do the letters BOD represent? [1]

    (ii) Explain what the BOD value, $350\,\mathrm{g\,m^{-3}}$ means. [2]

(b) Which effluent, A or B would contribute more to eutrophication? Give a reason for your answer. [2]

[Total: 5]

Chapter 9
Microbiology

The following topics are covered in this chapter:

- Diversity of microorganisms
- Microbial culture and measurement

9.1 Diversity of microorganisms

After studying this section you should be able to:

- describe the general characteristics of a range of important microorganisms
- understand how to use Gram staining to help identify bacteria

LEARNING SUMMARY

Some important microorganisms

AQA B	M7
EDEXCEL	M4
OCR	M5
WJEC	M4
NICCEA	M5

The microorganisms are members of kingdoms Prokaryotae, Protoctista and Fungi. Each type of organism has important structural differences.

Viruses

These are so simple in structure, that they are not considered members of any kingdom. They do not breathe, feed, excrete. They can however replicate.

The diagrams below show typical viral characteristics.

Rod shaped virus

A bacteriophage virus

Top view of bacteriophage virus

Many viruses are disease causing agents, e.g. the polio causing virus. Some viruses, e.g. the T2 bacteriophage can be helpful.

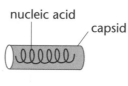

nucleic acid

capsid

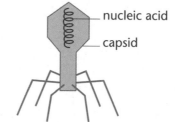

nucleic acid

capsid

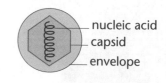

nucleic acid

capsid

envelope

Shape is variable but they do have some common features:

- an outer coat (**capsid**) consisting of protein units (**capsomeres**)
- an internal core of **RNA** or **DNA**
- they all **reproduce** by using the DNA of a **host cell**, so in this respect they are **parasitic**
- some viruses have an additional outer cover known as an **envelope**.

KEY POINT

The diagram shows the stages of viral attack on a host cell.

The bacteriophage can be used in genetic engineering to incorporate a gene into a host cell. On this occasion lysis would not take place. Instead a virus can be in **provirus** form. It is inactive and the host cell is **lysogenic**, in this state.

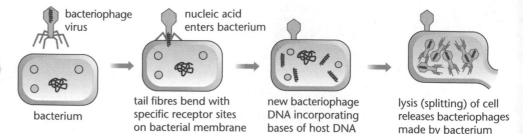

bacteriophage virus

nucleic acid enters bacterium

bacterium

tail fibres bend with specific receptor sites on bacterial membrane

new bacteriophage DNA incorporating bases of host DNA

lysis (splitting) of cell releases bacteriophages made by bacterium

Once they have replicated the viruses are ready to attack new host cells.

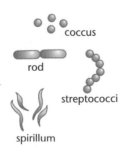

Retroviruses: useful or deadly?

Special consideration should be given to these viruses. They can be useful and they can be deadly! Here are the important features.

- They all have an RNA core.
- They all produce a special enzyme called **reverse transcriptase**.
- They are able to **synthesise a strand of DNA from a strand of RNA**.
- This is followed by the synthesis of the complementary DNA strand, so that a full **double stranded DNA can be formed**.
- The **DNA** forms a circular shape then **enters the host cell** and incorporates into the host DNA.
- Here it exists as a **provirus**, able to lie inactive for a number of years.
- However, the **host-viral DNA** is able to make **viral proteins**.
- Every time the host cell divides then so does the provirus, therefore the number of infected cells can replicate dramatically.

First the bad news. The above bullet points outline the action of the human immunodeficiency virus (HIV) which causes AIDS. The lymphocytes are attacked and the consequences, in time, are fatal.

Now the good news! Reverse transcriptase is used in genetic engineering. Imagine a genetic engineer is trying to locate a specific gene responsible for a particular protein, e.g. glucagon, along a chromosome. There may be a 1000 genes along the chromosome. Where does he or she begin to find the correct bases along a coding strand of DNA? Finding a needle in a haystack might be easier. Reverse transcriptase is an excellent tool for this process. The example below shows the principle.

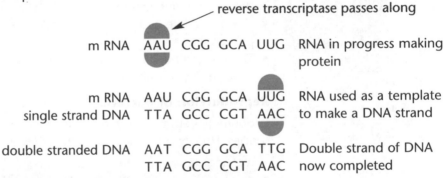

m RNA	AAU	CGG	GCA	UUG	RNA in progress making protein
				reverse transcriptase passes along	
m RNA	AAU	CGG	GCA	UUG	RNA used as a template
single strand DNA	TTA	GCC	CGT	AAC	to make a DNA strand
double stranded DNA	AAT	CGG	GCA	TTG	Double strand of DNA
	TTA	GCC	CGT	AAC	now completed

Bacteria

These are members of the kingdom Prokaryotae, and include saprobiotic and parasitic species. They exist in a number of different shapes. A selection is shown in the margin, together with a generalised structure.

coccus

rod

streptococci

spirillum

Typical bacterial features

- Cell wall which is not made of cellulose.
- No true nucleus, but the DNA is in nucleiod form, a single chromosome of coiled DNA, and in circular plasmids (in some bacteria).
- If flagellae are present there is not a 9 + 2 filament structure.
- Ribosomes are present but they are small.
- Usual reproduction by binary fission, a form of asexual reproduction.
- No membrane bound organelles.

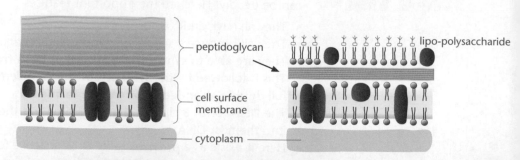

cell surface membrane
chromosome
cell wall
plasmid
food reserve granule
ribosomes
cytoplasm
slime capsule
pili
flagellae

Key structures labelled in red not always present.

Gram positive and Gram negative

Christian Gram devised his staining technique in 1884. It depends upon two different surface structures.

peptidoglycan
cell surface membrane
cytoplasm
lipo-polysaccharide

> **KEY POINT**
>
> Bacteria have either of two types of cell wall structure:
>
> - **Gram positive bacteria** – outside the cell surface membrane is a thick (around 8 nm) rigid layer of peptidoglycan.
> - **Gram negative bacteria** – outside the cell surface membrane is a thin (around 2 nm) layer of peptidoglycan, and additionally an extra outer membrane which includes lipopolysaccharides.

How can bacteria receive new DNA?

Conjugation – bacteria join by pili and can donate plasmid from one bacterium to another. Can be replicated each time bacterium divides.

Transformation – one bacterium donates DNA to another, which has now acquired new properties.

Transduction – a bacteriophage introduces DNA into a bacterium, which develops new properties.

What is the procedure for Gram's staining technique?

- Smear actively growing bacteria on a slide.
- Heat fix, then stain with crystal violet and dilute iodine.
- Wash slide off with ethanol or propanone.
- Gram positive bacteria retain the stain and show up as purple (the colour of the violet-iodine complex).
- Gram negative bacteria lose the stain as it is washed away with the ethanol or propanone.
- These Gram negative bacteria can be counterstained with a stain such as safranin O (red in colour).

Reproduction of bacteria

In optimum conditions reproduction can take place in around 20 minutes for some species. Usually by binary fission the following stages take place:

- the single chromosome replicates
- the chromosome at this time attaches to the cell surface membrane or a mesosome, which helps to part the two chromosomes
- a cross membrane and cross wall form in a central position, dividing off the two chromosomes, so that two daughter cells are produced.

Fungi

Organisms of this kingdom are eukaryotic and heterotrophic. In obtaining their complex organic substances they can be classified into parasitic, saprobiotic and mutualistic groups.

The diagrams below show typical fungal structure.

An aseptate fungus

typical fungal hyphae

A septate fungus

some fungi have septa (cross walls)

A single hypha

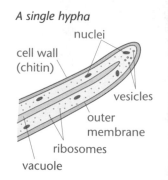

nuclei
cell wall (chitin)
vesicles
outer membrane
ribosomes
vacuole

Typical features of fungi

- Many have a thread-like basic unit known as a hypha.
- These hyphae are multinucleate.
- They have a cell wall of chitin, not cellulose.
- Nutrients are absorbed directly through the outer wall and membrane.
- They produce spores, either asexually or sexually.
- Bread mould (*Mucor hiemalis*) exemplifies all of the listed features.

Saccharomyces cerevisiae

This species of fungus is structurally different.

- This fungus is unicellular and is not multinucleate.
- It reproduces sexually or asexually by budding.

A single yeast cell

position where this cell detached from parent cell

cell wall (mostly mannose and glucose polymers)

mitochondria

cell surface membrane

rough endoplasmic reticulum

vacuole

Golgi body

cytoplasm

a bud beginning to form

Progress check

(a) Outline the procedure for the Gram staining technique.

(b) After staining how is it possible to identify Gram positive from Gram negative bacteria?

(b) Gram positive bacteria show up as purple but Gram negative do not.

(a) Smear (actively growing) bacteria on a microscope slide, heat fix, then stain with crystal violet and dilute iodine, wash slide off with ethanol or propanone.

9.2 Microbial culture and measurement

After studying this section you should be able to:

- *describe suitable conditions and nutrients for the growth of microorganisms*
- *describe techniques to culture microorganisms*
- *describe a number of techniques to measure microbial populations*

LEARNING SUMMARY

Nutritional needs

AQA B	M7
EDEXCEL	M4
OCR	M5
WJEC	M4

Before attempting to grow populations of microorganisms we need to know their nutritional requirements. These are needed as an energy source and for synthesis of cell components.

Key elements	Example of use
carbon, hydrogen, oxygen	a constituent of all organic molecules and structures
nitrogen	a constituent of all proteins
sulphur	a constituent of some proteins and co-enzymes
phosphorus	a constituent of phospholipids and ATP

- They also require potassium, magnesium, calcium and iron.
- All of the above elements are present in some form in culture media.
- Additionally, certain microorganisms need other specific substances.
- They do need **growth factors**, which are substances needed by the microorganisms because they cannot make them themselves.
- All substances must be present in balanced quantities to support the growth of the microorganisms.

Conditions for growth

Individual microorganisms have specific requirements. Some important requirements are listed below.

- Suitable temperature (Psychrophilic bacteria   optimum usually below 0 °C
  Mesophilic bacteria   optimum around 30 °C
  Thermophilic bacteria   optimum above 45 °C)
- Oxygen – some microorganisms are aerobic and some anaerobic. Yeast is a facultative anaerobe, having the ability to respire aerobically if oxygen is present.
- pH – most bacteria grow best at around pH 6.5 but will tolerate pH 4.0 – 9.0. *Thiobacillus thiooxidans* can grow at pH 3!

How can we prepare sterile agar plates?

1. sterile Petri dish — nutrient agar with water
2. autoclave at 121°C (pressure build up elevates boiling temperature, better for sterilisation)
3. sterilise flask top with flame
4. Pour the plate — lid prevents microbe entry from air
5. allow to set (can be stored in the refrigerator) — Petri dish

Nutrient broth (a liquid medium) is also used to culture microorganisms. Both agar and broth are examples of nutrient media. Often a broad spectrum medium is used, suitable for a wide range of microorganisms. Usually a buffer is present in the medium, to maintain the pH.

The conditions in which a plate is incubated are selective. In school, the incubation temperature for plates is normally 25°C. This is to avoid pathogens which attack humans. They would be encouraged to grow if a temperature of 37°C was used.

Can nutrient media be selective?

> It is possible to use a highly selective growth medium, suitable for only a specific organism. This medium will not support the growth of other microorganisms even if they are present. They are inhibited.

KEY POINT

Example 1

Culture of *Azotobacter* (nitrogen fixing bacteria found in soil).

- The medium used should contain sugar and essential minerals.
- **No ammonium ions** would be added because *Azotobacter* obtain their nitrogen from atmospheric nitrogen.
- Initially the **only bacteria growing** would be *Azotobacter* with **other bacteria suppressed.**
- **Later,** as *Azotobacter* die, they release NH_4^+, so that **other microorganisms can develop.**

Example 2

Culture of *Salmonella* (pathogenic bacteria found in human gut).

- MacConkey agar includes sodium taurocholate (a bile salt which **prevents growth of Gram positive bacteria**) and lactose.
- It contains a **pH indicator** which **changes to red if the pH drops to less than 6.8.**
- Other bacteria, e.g. *E. coli*, use lactose and as a result produce an acidic waste product, so with the indicator present, these bacteria show up as red.

- *Salmonella* **bacteria** do not use the lactose and **remain colourless**.

So the red colonies are other bacteria, Gram positive bacteria do not grow, and the colourless colonies are the dangerous *Salmonella*. This would be excellent diagnostic information for doctors!

Streaking a plate

The term plate refers to a Petri dish. The diagrams below show how a sterile inoculating loop is used to pick up and transfer microorganisms.

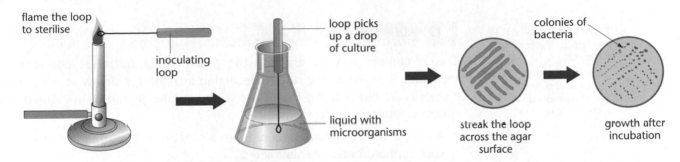

flame the loop to sterilise

inoculating loop

loop picks up a drop of culture

liquid with microorganisms

streak the loop across the agar surface

colonies of bacteria

growth after incubation

Other techniques can be used but all parts of the process must be **aseptic**. Any transfer instrument must be sterilised in some way.

Taping a Petri dish ready for incubation

Taping is important. The lid should be kept on at all times but air is still allowed into the dish. This encourages aerobes and discourages dangerous anaerobes from developing.

Dilution plating

In one Petri dish there can be millions of bacteria which are impossible to count accurately. A dilution technique must be employed.

> **Remember this principle.** One **viable**, bacterium, when transferred to a suitable medium, will reproduce to form a circular colony. One bacterium is invisible to the naked eye but we work backwards. If there are 20 colonies which have grown then we know that we began with 20 individual bacteria on day 1.
>
> **KEY POINT**

Pond water and milk contain too many bacteria to be directly counted. When plated the colonies of bacteria would merge together. Any count would be inaccurate!

The dilution technique

- Take a 1 cm^3 sample of say, pond water, then add 9 cm^3 sterilised water.
- The population of microorganisms is now at 1 in 10 cm^3 of the initial sample.
- Mix thoroughly then take 1 cm^3 of the diluted population and add 9 cm^3 sterilised water.
- The population of microorganisms is now at 1 in 100 or 10^{-2} of the initial sample.
- The above dilution process can be continued down to 1 in 10 000.
- The 1 in 10 000 (10^{-4}), 1 in 1000 (10^{-3}), 1 in 100 (10^{-2}), samples are used and for each, 1cm^3 is transferred to molten agar which is added to Petri dishes, which are incubated for 48 hours.
- Counts are then made, colony numbers in a plate from 30 – 300 are counted.
- If there are less than 30 colonies, the number is considered not reliable.
- If there are more than 300 colonies, the colonies merge with each other so that reliable counting is not possible.
- Remember that every circular colony can be tracked back to a single bacterium!

The principle of dilution plating.

A worked example

If you found that the plates for a 1 in 10000 dilution were below 30 then they would be immediately discounted.

At a dilution of 1 in 1000 (10^{-3}) a sample of $0.2\,cm^3$ is plated and the average plate count is 96.5 colonies.

Number of colonies in $1\,cm^3$ of 10^{-3} dilution pond water $= \dfrac{96.5 \times 1}{0.2} = \mathbf{482.5}$

Number of bacteria in $1\,cm^3$ original pond water $= \dfrac{482.5 \times 1}{10^{-3}} = \mathbf{482\,500}$

Turbidimetry is not as accurate as dilution plating as:

- the cloudiness or opacity of the culture is measured – dead microorganisms as well as the viable ones
- it does not give numbers of microorganisms.

Turbidimetry

This is another way to estimate the growth in a microbial population. The technique is based on the culture becoming increasingly cloudy as the population increases. Turbidity is the cloudiness of the culture. It can be measured using a spectrophotometer or colorimeter.

- A sample of the culture is poured into a cuvette and inserted into the spectrophotometer or colorimeter.
- Light is passed through each sample.
- The amount of light received by a sensor after being passed through the sample is indicated via a meter.
- The less light reaching the sensor, the greater the growth of the microbial population.
- The more light absorbed by the culture, the greater the growth of the microbial population.

Sample question and model answer

(a) The diagram below shows a Petri dish containing a **selective** agar medium for **nitrogen fixing bacteria**. The agar had all the essential nutrients but ammonium ions were excluded. The diagrams show the growth of bacteria after 2 days, 8 days and 14 days. Colonies of species X were visible after two days and an additional species, Y, appeared after 14 days.

after 2 days   after 8 days   after 14 days

bacteria species X   bacteria species Y

This question illustrates an interesting point. Bacteria Y appeared after 14 days but they must have been present at the beginning. They were just too small to see!

(i) Why did **only** species X grow during the first few days of the investigation? [4]

No ammonium ions added so species Y could not grow.

Species X bacteria obtain their nitrogen from atmospheric nitrogen, so can make their proteins, having all requirements for growth.

Species X releases ammonium (NH_4^+), so that species Y can now grow.

(ii) How was the agar medium inoculated with the bacteria? Give evidence from the diagrams to support your answer. [1]

Streaking, because the colonies are growing in lines.

(iii) Why were circular colonies of bacteria not visible after 14 days? [1]

The colonies had grown together or merged together.

(b) After two days there were 39 colonies of species X. Assuming,

(i) the original sample containing the bacteria was diluted to 1 in 10 000

(ii) $0.1\,cm^3$ of the diluted sample was plated to produce the 39 colonies, how many **viable** bacteria were there in $1\,cm^3$ of the original sample? Show your working. [2]

There are often mathematics questions in biology papers. Always practise before you take your 'live' examination. This Guide will help you.

In $1\,cm^3$ of the 1 in 10 000 dilution there were $\dfrac{39 \times 1}{0.1} = 390$

But the dilution factor is 1 in 10 000 or 10^{-4}

so number of viable cells is $\dfrac{390}{10^{-4}} = 3\,900\,000$ in $1\,cm^3$ of the original sample.

(c) If you wished to measure population growth and include dead bacteria as well as the living ones which method would you use? [1]

Turbidimetry, using a spectrophotometer or colorimeter.

Practice examination questions

1 (a) The table below includes features shown by viruses, bacteria and fungi. Complete the table below by putting a tick in each correct box, for a feature shown by the group of organisms.

	Virus	Bacterium	Fungus
has membrane bound organelles			
has an outer protein coat of capsomeres			
has ribosomes			
is prokaryotic			
is multinucleate			
cannot respire			
has plasmids			
has an outer wall of chitin			
has nucleic acid but no cytoplasm			

[9]

(b) The term bacteriophage can link two of the above groups of organisms.

Describe the relationship which links the two groups together. [2]

[Total: 11]

2 A retrovirus made double stranded DNA from mRNA which was in the process of protein synthesis. The base sequence below shows part of the first single strand of DNA which the virus was able to synthesise.

GGC TTA ATC GCT AAG TAC   single strand DNA

(a) (i) What sequence of bases on the mRNA strand would have coded for the production of the first single strand of DNA? [1]

(ii) What is the complementary strand to the first single strand of DNA? [1]

(b) Which type of enzyme would the virus use to synthesise the double stranded DNA form the mRNA? [1]

[Total: 3]

3 A patient submits a sample of faeces to the pathology department of a hospital.

(a) Outline how the laboratory technician would test for the presence of suspected *Salmonella* bacteria, using MacConkey agar. [2]

(b) Explain how the MacConkey agar allows *Salmonella* to be differentiated from other Gram negative bacteria. [4]

(c) Explain what happens to Gram positive bacteria if they are present in the sample transferred to the plate. [2]

[Total: 8]

Chapter 10

Biotechnology

The following topics are covered in this chapter:

- *Large scale production*
- *Medical applications*

- *Further gene transfer*

10.1 Large scale production

After studying this section you should be able to:

- *describe the main features of batch and continuous culture*
- *outline fermentation processes which yield substances useful to human requirements*
- *describe the production of mycoprotein using a commercial fermenter*
- *describe the production of yoghurt and yeast extract*

LEARNING SUMMARY

An introduction to biotechnology

AQA B	M7
EDEXCEL	M4
OCR	M5
WJEC	M4

Before describing how substances are commercially produced it is necessary to consider the meaning of the term, **biotechnology**.

> Biotechnology is the use of organisms and biological processes to supply nutrients, other substances and services to meet human needs.
> **Fermentation** is a key process in biotechnology as microorganisms are used to produce traditional products such as ethanol and the more recent production of substances such as pharmaceutical chemicals and the enzymes for biological washing powder.
>
> **KEY POINT**

Batch fermentation

In commercial fermentation large scale fermenters are used in production. Batch fermentation takes place in a **closed** vessel.

> - The fermenter is steam sterilised, then sterile nutrients are added.
> - Fermentation commences and optimal conditions are maintained throughout the process.
> - Fermentation continues until a maximum level of product is reached, when the process is stopped and the yield of product harvested.
> - Harvesting time is critical because waste substances can result in a decrease of the product.
> - So harvesting must be implemented before this takes place.
>
> **KEY POINT**

Advantages

- If the culture becomes contaminated in any way, just one batch is spoiled.
- The fermenter can be used for a variety of fermentation processes, e.g. different antibiotics.

Disadvantages

- At the end of every production period **shut down** takes place. The vessel needs to be cleaned and re-sterilised. This lost time can be expensive to the company.
- Often the product, waste substances and unused nutrients are mixed together, e.g. in penicillin production. Product removal is made more difficult by these contaminants.

Key points from AS

- **Modern industrial fermenters**
  Revise AS page 88

Continuous fermentation

Continuous fermentation takes place in an open fermenter.

- The fermenter is steam sterilised.
- Regular amounts of sterile nutrients are added.
- At the same time regular amounts of product are removed.
- Optimum levels of pH, oxygen, nutrients and temperature are maintained.

Advantages

- The rate of growth of the microbial population is kept at a maximum level: this is known as the **exponential rate**.
- There is **no** regular pattern of **shut down**.

Disadvantage

- Maintaining the levels at **optimal levels** is difficult.
- **Regular** sampling is necessary for quality control, ensuring that chemicals are in equilibrium, and contaminants are absent.
- There is more chance of contaminants entering due to regular input and output.

The continuous production of mycoprotein in an air-lift fermenter

No stirrer required! Mixing is by the air bubbles produced from the sparger. A stirrer would break the hyphae which are needed to give a meat-like texture to the product.

gas outflow valve (mainly CO_2)

sparger

NH_3 + air in

glucose, K^+, Mg^{2+}, PO_4^-

harvest line (mycoprotein out, heat shocked, dried)

cooling control system

- *Fusarium graminearum* is the fungus which is cultured to produce mycoprotein.
- The fermenter is operated continuously for six weeks during which there is a steady input of nutrients and an equal output of product.
- Air pumped into the fermenter supplies the required oxygen for the aerobic fungus and agitates (mixes!) the culture.
- Gas outlet of carbon dioxide takes place at the top of the fermenter.
- Probes monitor internal conditions by interfacing with a computer so that modifications to the internal environment can be made during fermentation (optimum temperature is at 30°C).
- Fungal hyphae can be filtered out during the harvesting process.
- The harvested fungus is then heat shocked at 65°C which denatures the fungal nuclei.
- Addition of flavours converts the very healthy, low fat food into a popular food sold in all major supermarkets as Quorn™.

KEY POINT

Commercial production of cheese

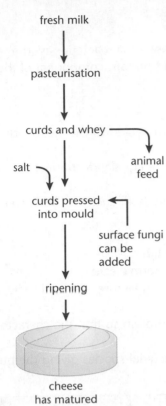

fresh milk
↓
pasteurisation
↓
curds and whey ⟶ animal feed
↓
salt ⟶ curds pressed into mould ← surface fungi can be added
↓
ripening
↓
cheese
has matured

The process

- Milk is first **pasteurised** to destroy unwanted microorganisms.
- Selected bacteria are added such as:
  - *Leuconostoc lactis* which **produces CO$_2$** and has a major role in the determination of cheese **texture**
  - *Streptococcus lactis* which produces **lactic acid** from the milk sugar, lactose, and has a major role in the determination of **flavour**.
- Lactic acid **coagulates** (clots) the milk protein (casein) to form solid curds in a liquid component, the whey.
- **Rennet** is added which also coagulates the milk protein. (The rennet is usually artificially produced from genetically modified bacteria in fermenters.)
- The whey is removed and used in pig food.
- The curds have salt added as an osmotic preservative and they are pressed into a mould.
- In the coming weeks and months the bacteria and often fungi continue to grow on the surface and through the cheese.
- Enzymically driven changes take place such as protein breakdown to amino acids, until finally a characteristic cheese taste is achieved after this **ripening**.

Commercial production of beer

Beer is produced by the fermentation of a barley/hop mixture using a strain of *Saccharomyces cerevisiae* (yeast).

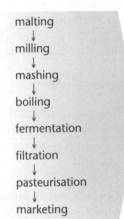

malting
↓
milling
↓
mashing
↓
boiling
↓
fermentation
↓
filtration
↓
pasteurisation
↓
marketing

The process

- **Malting** – germinating the barley so that amylase is produced.
- **Heating** to 80 °C to **kill the seed embryos** but **not denature amylase**.
- **Milling** – the malted barley is ground into a powder called grist (increase in surface area!).
- **Mashing** – hot water addition plus amylase, hydrolyses starch into wort.
- **Boiling** – hops added to wort at high temperature which imparts flavour.
- **Fermentation** – the addition of *S. cerevisiae* results in ethanol production.
- **Finings** – are added to clear the beer.
- **Filtration** – this follows the addition of finings to clear the beer, removes any yeast cell debris.
- **Pasteurisation** – to prevent any further fermentation in the marketed container.

Meat tenderisation

Traditionally a carcass of a meat animal such as a bullock is hung for several days, before the beef is sold to the public, during which proteolytic enzymes hydrolyse the proteins. This tenderises the meat.

The modern technique is to use enzymes which make the meat tender, e.g. papain from the papaya plant.

Advantages

The process is helpful to the people selling the product rather than the consumer!

- Storage time of meat is reduced since tenderisation is fast.
- Older animals with tough meat, not normally marketed as a prime product, can yield a better profit due to the tenderisation treatment.

Commercial production of yoghurt

Yoghurt is made by fermentation using a range of bacteria as the active microorganisms and milk as a food substrate. An important component of the milk is lactose, milk sugar.

Key points of the process

- Milk is heat treated at 90 °C for 30 minutes then cooled to the incubation temperature.
- The milk is homogenised so that components are equally distributed throughout.
- The fermenter is inoculated with a starter culture (e.g. *Lactobacillus bulgaricus* and *Streptococcus thermophilus*).
- Incubated at around 40 °C for about 6 hours.
- *L. bulgaricus* breaks down milk protein into peptides.
- *S. thermophilus* uses the peptides and produces formic acid.
- *L. bulgaricus* uses the formic acid and breaks down lactose into lactic acid, so the pH falls to about 4.3.
- Lactic acid also causes the coagulation of milk protein so that the characteristic thickening of the yoghurt takes place.
- Ethanal (acetaldehyde) is produced by each bacterial species giving the buttery taste, whereas lactic acid gives a 'tangy' taste.

> **Microbial processes**
>
> Note that there are common parts to each process involving microorganisms. Pre-process sterilisation is always carried out with steam. Disinfectant would contaminate the product. A suitable temperature is always required so that the microbial enzymes function at an optimum rate. The pH level is always monitored carefully.
>
> **KEY POINT**

Yeast extract: a by-product of 'spent' brewers' yeast

Yeast is produced during the production of ethanol. During fermentation as the ethanol increases then so does the yeast population, yielding large quantities of a potentially useful by-product. Yeast contains the full range of B complex vitamins.

Additionally a **hydrolysate** can be produced:

- heat yeast with HCl to change proteins to amino acids
- neutralise with NaOH
- tastes meaty and salty.

Salt is produced as the HCl is neutralised by NaOH.

How is the yeast processed?

- Yeast is warmed to 50°C which results in **self-digestion** known as **autolysis**.
- Carbohydrates, proteins and lipids are hydrolysed by enzymes within the yeast cells.
- Breakdown products such as amino acids and glycerol form in the organic mixture.
- These substances are known as the **autolysate**.
- Separation of these useful substances is by **filtration** and **centrifugation**.
- After dehydration it can be used as a food additive to impart a meaty flavour.

Progress check

State **two** differences between continuous and batch production.

- Batch production is shut down on a more regular basis.
- In continuous production nutrients are added on a regular basis, and products are removed in similar quantities. In batch production product retrieval is at the end rather than during the process.

10.2 Medical applications

After studying this section you should be able to:

- *outline the production of antibiotics*
- *describe the production and applications of monoclonal antibodies*

Antibiotic production

AQA B	M7
EDEXCEL	M4
OCR	M5
WJEC	M4

Antibiotics are substances produced by microorganisms which kill or inhibit further growth of other microorganisms. There are a number of different antibiotics, including penicillin.

- Discovery of penicillin was by Sir Alexander Fleming in 1928.
- He grew the bacterium *Staphylococcus* on agar.
- *Penicillium notatum* spores had reached the agar and had begun to grow.
- Next to the fungal mycelium a gap remained where no bacteria would grow.
- He found that a substance had been secreted by the fungus which he named penicillin.
- He followed this up with further research to show that the penicillin killed a number of pathogenic organisms.

> Note that the fermenter on p.88 in the AS guide would be used for penicillin production.

KEY POINT

Antibiotics can be either bactericidal or bacteriostatic in action.

Bactericidal is the term used when they kill the microbe which they attack, e.g. penicillin is bactericidal.

Bacteriostatic is the term used when antibiotics halt further microbial growth.

Penicillin production

> Penicillin is able to destroy a narrow range of bacteria so is classified as a **narrow spectrum antibiotic**.
>
> Chloramphenicol can destroy a wide range of bacteria and so is classified as a **broad spectrum antibiotic**.

- Large scale production takes place in a fermenter.
- The nutrient medium usually contains glucose and lactose.
- The *Penicillium* fungus is usually cultured in a fermenter for about 6–8 days.
- Production is by batch culture so that harvesting comes when the penicillin yield is considered to be at a maximum (see graph below).
- The fungal hyphae can be removed by filtration.
- The remaining liquid contains the antibiotic.
- K^+ ions are added to this liquid component forming a compound with the penicillin.
- This compound precipitates so that it can readily be removed.
- After drying, the product is removed at over 99% purity.

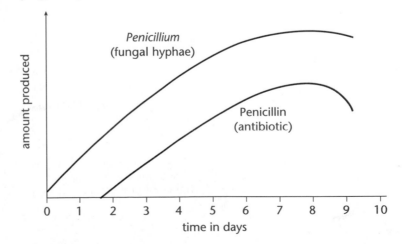

Monoclonal antibodies

AQA B	M8
EDEXCEL	M5
OCR	M5
WJEC	M4

Antibodies are produced by B-lymphocytes in response to specific antigens. We have many different lines of B-lymphocyte, each being so specific in action that it will only bind with a highly specific antigen. Many researchers looked towards production of antibodies, targeted at cancer cells.

It is hoped that these will be the answer: **monoclonal antibodies**.

Key points from AS

• **How do antibodies destroy pathogens**
Revise AS page 136

The problems

• Isolated lymphocytes die quickly outside an organism.
• More robust cells are needed to exist in production vessels outside the organism if commercial antibody production is to be possible.

Principles of production

• An organism, e.g. mouse is injected with antigen, e.g. red blood cells of a sheep.
• Antigens on the red cells stimulate B-lymphocytes to produce antibodies.
• The B lymphocytes produce plasma cells which are removed from the spleen of the mouse.
• Tumour cells from the bone marrow of another mouse are collected.
• Plasma cells are fused with tumour cells forming a **hybridoma**, a very robust cell.
• The resulting cells are incubated on a special medium; only cells which have successfully fused and become hybridomas can survive.
• Hybridoma cells are then cloned on media in a laboratory.
• The hybridoma cells are assayed quantitatively for antibody secretion and only the best line of hybridoma cells is selected.
• Large scale production is in an air-lift fermenter to prevent damage to the cells.

Do not get confused! At the end of this process the antibodies could **only** be targeted at the antigens on the red blood cells.

Whichever antigen is injected into the first animal is the one which the antibody will successfully bind to.

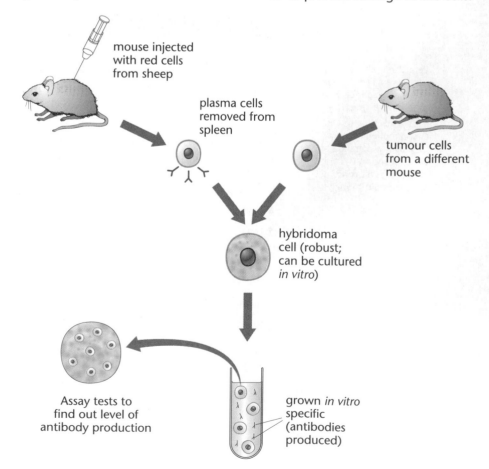

mouse injected with red cells from sheep

plasma cells removed from spleen

tumour cells from a different mouse

hybridoma cell (robust; can be cultured *in vitro*)

Assay tests to find out level of antibody production

grown *in vitro* specific (antibodies produced)

What can monoclonal antibodies be used for?

Treatment of tetanus – the tetanus antigen is injected into a mouse, as seen in the process above. The antibodies produced are reasonably successful but, owing to their 'mouse' origin they are recognised as antigenic themselves and eventually attacked by the immune system.

Monoclonal antibodies and pregnancy testing

The principles

- The hormone **human chorionic gonadotrophin (HCG)** is present in the urine of a pregnant woman.
- The base of the pregnancy testing strip is put into urine.
- At the base are **coloured mobile monoclonal antibodies** which have been produced **against HCG**.
- If HCG is present in the urine then these antibodies bind to the HCG and move up the test strip by capillary action.
- Midway up the strip is a line of **immobilised antibodies** capable of binding to HCG.
- Resulting along this **mid-line** is a complex of three molecules:

    mobile antibodies + HCG + immobilised antibodies

  which show up as a **blue line**, a **positive pregnancy test**.
- In a negative test, HCG would not be bound to the mobile antibodies.
- The mobile antibodies then rise to the top of the testing strip where they bind to an immobilised line of complementary antibodies.
- The antibodies bind together to give a **blue line at the top** to show that the **woman is not pregnant**.

Not all of the mobile antibodies bind with HCG, even though it is present in the urine sample. The result is that some mobile antibodies rise to the top of the strip, bind with the immobilised antibodies there producing another blue line. TWO blue lines show up for a pregnant woman!

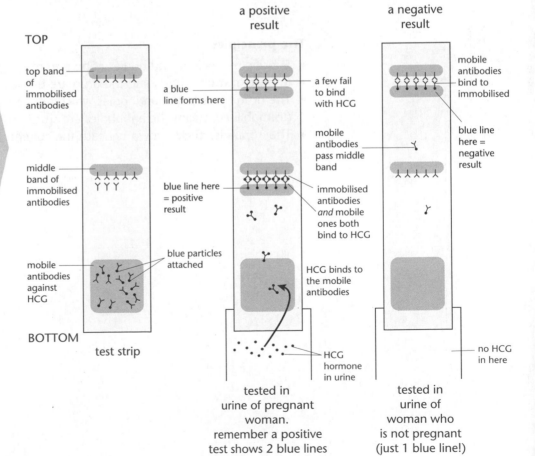

test strip

a positive result

a negative result

tested in urine of pregnant woman.
remember a positive test shows 2 blue lines

tested in urine of woman who is not pregnant (just 1 blue line!)

How can monoclonal antibodies be used to purify drugs?

It is vital to produce drugs such as interferon at a very high level of purity. This can be achieved with monoclonal antibodies in a resin-based column.

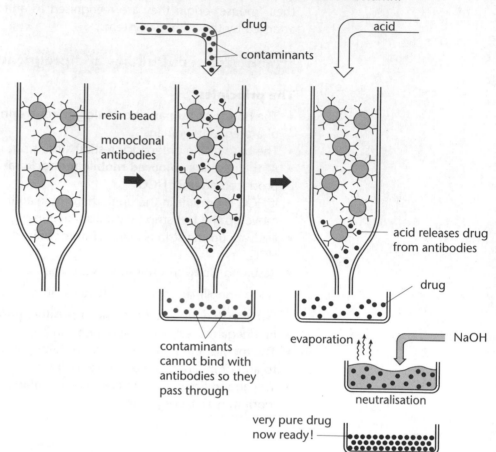

The principles

- Specific monoclonal antibodies adhere to resin beads in a glass column.
- Impure mixtures containing the drug are trickled through the column.
- The drug binds to the antibodies which are immobilised on the beads. (Immobilised means the antibodies are stuck on to the beads!)
- The impurities trickle down beneath the column and are removed.

10.3 Further gene transfer

After studying this section you should be able to:

* describe a range of transgenic applications
* describe advantages and potential disadvantages of genetic modification

Transgenic organisms

EDEXCEL	MS
OCR	MS
WJEC	M4

Modern gene technology will result in huge advances during the shelf-life of this book. Moving genes from one species to another may have great benefits.

It will be necessary to balance these benefits against the potential dangers and the ethics of interfering with the genome of a species.

> Transgenic organisms have very useful genes, acquired by **gene transfer from other organisms**. The stage at which the gene transfer is made is critical. This is usually when the gene(s) insertion takes place into the **nucleus of a zygote**, before the first mitotic division. Every subsequent cell division passes on this gene to every somatic cell of the organism. Additionally the new gene will be passed on to offspring; the gene has been incorporated into the genome of the species.

KEY POINT

Transgenic crop plants

Inserting genes into crop plants is becoming increasingly important in meeting the needs of a rising world population. A range of techniques is used to engineer new genes into a species.

In **plants** there is an important technique which uses a **vector** to insert a novel gene, the bacterium *Agrobacterium tumefaciens*.

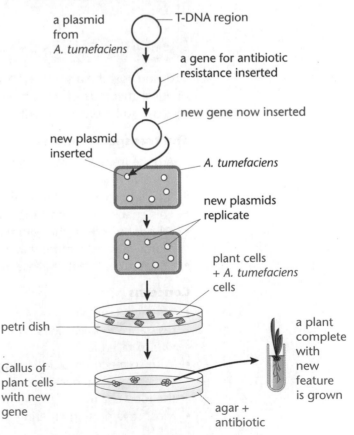

a plasmid from *A. tumefaciens* — T-DNA region

a gene for antibiotic resistance inserted

new gene now inserted

new plasmid inserted

A. tumefaciens

new plasmids replicate

plant cells + *A. tumefaciens* cells

petri dish

Callus of plant cells with new gene

agar + antibiotic

a plant complete with new feature is grown

Key points from AS

* **Genetically modified organisms**
  Revise AS page 90
* **Gene technology**
  Revise AS page 87

The principle of using *A. tumefaciens* can be used in gene transfer to many different plants. Applications are in an early stage of development. Look to the media!

KEY POINT

Agrobacterium tumefaciens

- This is a **pathogenic bacterium** which **invades** plants forming a gall (abnormal growth).
- The bacterium contains **plasmids** (circles of DNA) which carry a gene that stimulates tumour formation in the plants it attacks.
- The part of the plasmid which does this is known as the **T-DNA region** and can insert into any of the chromosomes of a host plant cell.
- Part of the T-DNA controls the production of two growth hormones, auxin and cytokinin.
- The extra quantities of these hormones stimulate rapid cell division, the cause of the tumour.

How can *Agrobacterium tumefaciens* be used in gene transfer?

KEY POINT

- Firstly, the DNA section controlling auxin and cytokinin was deleted, tumours were not formed, and cells of the plant retained their normal characteristics.
- A gene which gave the bacterial cell **resistance to a specific antibiotic** was inserted into the T-DNA position.
- The **useful gene** was **inserted into a plasmid**.
- Plant cells, minus cell walls, were removed and put into a Petri dish with nutrients and *A. tumefaciens*, which contained the engineered plasmids.
- The cells were **incubated** for several days then transferred to another Petri dish containing nutrients plus the specific antibiotic.
- **Only plant cells with antibiotic resistance and the desired gene grew.**
- Any surviving cells grew into a callus, from which an adult plant formed, complete with the transferred gene.

Genetically modified soya bean plants – case study 1

Growing soya beans in the USA was big business and a big problem! Inefficient, physical methods of killing weeds had to be used because the crop plant was sensitive to herbicide (weedkiller).

The transgenic answer

- A gene from a plant resistant to herbicide was transferred into a soya bean plant.
- A new line was bred from this first resistant variant.
- This enabled the herbicide, **glyphosate**, to be used in the soya fields.
- Without weeds in the field there is no competition from weeds, so more light, more water, and more mineral ions are available to the crop.
- The result: a great increase in yield!

Concerns

- The new soya plants may interbreed with weeds around the fields and pass on resistance to herbicides. What effect would the herbicide have on weeds if this was to take place?
- With so much herbicide being sprayed in the field, what would happen if some of the chemicals remained in the food so that it passed to the humans in the food chain?
- Most soya protein entering the UK is of this genetically modified type.

Bovine somatotrophin (BST) – case study 2

Cattle produce a growth hormone known as **bovine somatotrophin (BST)** from their pituitary gland. The use of BST can improve the rate of meat production so that the animals are ready for the market earlier and have very lean meat, owing to their younger age. Additionally an increased milk yield, around 20%, is possible.

Scientists attempted to find a source of BST to inject into cows.

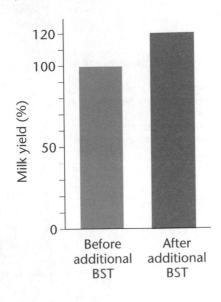

The transgenic answer

- The gene for BST production was transferred from cow cells into bacteria.
- The usual technique of gene transfer using restriction endonuclease was followed and the gene inserted into bacteria.
- The transgenic bacteria are grown in industrial fermenters where they secrete BST.
- After injection with BST cows produce more milk.

Concerns

- Injecting a healthy cow on a regular basis may be considered cruel.
- The number of cows becoming infertile is greater in those being injected with BST.

BST in milk will be consumed by people, but it has no effect because BST, as a protein, is digested in the stomach.

The Flavr Savr™ Tomato – case study 3

As tomatoes mature they change from green to red and become softer. Often they are picked green and turn red during transit. The taste is not good after this treatment. They can go soft very quickly and have a short shelf-life.

Flavr Savr™

The transgenic answer

- The tomatoes ripen due to the conversion of amino acids into ethene.
- Ethene stimulates the production of pectinase enzymes which break down the middle lamellae between tomato cells, softening them.
- Two genes have been transferred into a tomato which inhibit both changes.
- This results in the Flavr Savr™ tomato staying firmer for longer; the shelf-life has improved!

Sample question and model answer

The diagrams below show the transfer of a useful gene from a donor plant cell to the production of a transgenic crop plant. The numbers on the diagram show the stages in the process.

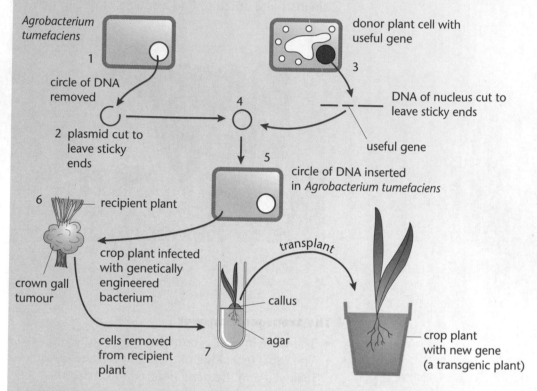

Agrobacterium tumefaciens

1

circle of DNA removed

2 plasmid cut to leave sticky ends

4

donor plant cell with useful gene

3

DNA of nucleus cut to leave sticky ends

useful gene

5

circle of DNA inserted in *Agrobacterium tumefaciens*

6 recipient plant

crown gall tumour

crop plant infected with genetically engineered bacterium

cells removed from recipient plant

7

callus

agar

transplant

crop plant with new gene (a transgenic plant)

(a) Give the correct name for the circle of DNA found in the bacterium, *A. tumefaciens*. [1]

plasmid

(b) The same enzyme was used to cut the DNA of the bacterium and of the plant cell.

(i) Name the type of enzyme used to cut the DNA. [1]

restriction endonuclease

(ii) Explain why it is important to use exactly the same enzyme at this stage. [2]

The same enzyme produces the same sticky ends.

Complementary sticky ends on donor gene bind with the sticky ends of the plasmid.

(iii) Which type of enzyme would be used to splice the new gene into the circle of DNA? [1]

ligase

(c) How was the new gene incorporated into the DNA of the crop plant cells? [2]

Crop plant infected by genetically engineered bacterium.

The DNA of bacterium causes a change in the DNA of crop plant to produce the gall or tumour cells.

(d) How would you know if the gene had been transferred successfully? [1]

The feature would be expressed in the transgenic plants.

Practice examination questions

1 The diagram below shows an industrial fermenter used to produce the antibiotic, penicillin.

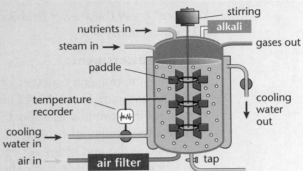

stirring
nutrients in →
alkali
steam in →
gases out
paddle
temperature recorder
cooling water out
cooling water in →
air in →
air filter
tap

(a) Describe **three** ways in which aseptic conditions are achieved in the fermenter. [3]

(b) If the air filter failed, explain what would be the likely effect inside the fermenter. [3]

(c) Which type of culture, batch or continuous, is used to produce penicillin? Give a reason for your answer. [2]

[Total: 8]

2 The statements below describe the principles of production of monoclonal antibodies. Place the statements in the correct sequence.

A Antigens on the red cells stimulated B-lymphocytes to produce antibodies

B Hybridoma cells are assessed for antibody secretion and only the best line of hybridoma cells is selected

C Plasma cells are fused with tumour cells forming hybridoma cells

D The resulting cells are incubated on a special medium, on which only hybridomas can survive

E B-lymphocytes produce plasma cells which were removed from the spleen of the mouse

F Hybridoma cells are cloned on media in a laboratory

G Tumour cells from the bone marrow of a different mouse are collected

H A mouse was injected with red blood cells [8]

3 The graph below shows the level of product secreted by microorganisms in a commercial fermenter.

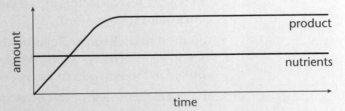

amount

product

nutrients

time

(a) Account for the shape of the graph. [1]

(b) Which type of culture, batch or continuous, took place in this fermenter? Give **two** reasons for your choice. [2]

[Total: 3]

Chapter 11
Synoptic assessment

What is synoptic assessment?

You must know the answer to this question if you are to be fully prepared for your A2 examinations!

Synoptic assessment:

- involves the drawing together of knowledge, understanding and skills learned in different parts of the AS/A2 Biology courses
- requires that candidates apply their knowledge of a number of areas of the course to a variety of contexts
- is tested at the end of the A2 course by both assessment of investigative/ practical skills and by examinations
- is valued at 20% of the marks of the course total.

Each Examination Group identifies which parts of its specification will be tested in the end-assessed synoptic questions. (Synoptically assessed content is highlighted in the specification grids for each Examination group in the specification list section of this book.)

Practical investigations

You will need to apply knowledge and understanding of the concepts and principles, learned throughout the course, in the **planning**, **execution**, **analysis** and **evaluation** of each investigation.

How can I prepare for the synoptic questions?

Why are synoptic skills examined?

Once studying at a higher level or in employment, having a narrow view, or superficial knowledge of a problem, limits your ability to contribute. Having discrete knowledge is not sufficient. You need to have confidence in applying your skills and knowledge.

- **Check out the modules** which will be examined for your specification's synoptic questions.
- Expect **new contexts** which draw together lots of different ideas.
- Get ready to **apply** your knowledge to a new situation; contexts change but the **principles remain the same**.
- In modular courses there is sometimes a tendency for candidates to learn for a module, achieve success, then forget the concepts. Do not allow this to happen! **Transfer concepts** from one lesson to another and from one module to another. Make those connections!
- Improve your powers of analysis – **take a range of different factors into consideration** when making conclusions; synoptic questions often involve both graphical data and comprehension passages.
- Less able candidates make limited conclusions; high ability candidates are able to consider several factors at the same time, then make a **number of sound conclusions** (not guesses!).
- You need to do **Regular Revision** through the course; this keeps the concepts 'hot' in your memory, 'simmering and distilling', ready to be **retrieved** and **applied** in the synoptic contexts.
- The bullet point style of this book will help a lot; back this up by summarising points yourself as you make notes.

148

Synoptic favourites

The final modules, specified by the Examination Group for synoptic assessment, include targeted synoptic questions. Concepts and principles from earlier modules will be tested together with those of the final modules. You can easily identify these questions, as they will be longer and span wide-ranging ideas.

Can we predict what may be regularly examined in synoptic questions?

'Yes we can!' Below are the top five concepts. Look out for common processes which permeate through the other modules. An earlier module will include centrally important concepts which are important to your understanding of the rest.

KEY POINT

Check out the synoptic charts!

1 Energy release

Both aerobic and anaerobic respiration release energy for many cell processes. Any process which harnesses this energy makes a link.

Examples

- Reabsorption of glucose involves active transport in the proximal tubule of a kidney nephron. If you are given a diagram of tubule cells which show both mitochondria and cell surface membrane with transporter proteins, then this is a cue that active transport will probably be required in your answer.
- Contraction of striated (skeletal) muscle requires energy input. This is another link with energy release by mitochondria and could be integrated into a synoptic question.
- The role of the molecule ATP as an energy carrier and its use in the liberation of energy in a range of cellular activity may be regularly linked into synoptic questions. The liberation of energy by ATP hydrolysis to fund the sodium pump action in the axon of a neurone.
- The maintenance of proton gradients by proton pumps are driven by electron energy. Any process involving a proton pump can be integrated into a synoptic question.

2 Energy capture

Photosynthesis is responsible for availability of most organic substances entering ecosystems. It is not surprising that examiners may explore knowledge of this process and your ability to apply it to ecological scenarios.

Examples

- Given the data of the interacting species in an ecosystem you may be given a short question about the mechanism of photosynthesis then have to follow the energy transfer routes through food webs.
- Often both photosynthesis and respiration are examined in a synoptic type question. There are similarities in both the thylakoid membranes in chloroplasts and cristae of mitochondria.
- Many graphs in ecologically based questions show the increase in herbivore numbers, followed by a corresponding carnivore increase. Missed off the graph, your knowledge of a photosynthetic flush which stimulates herbivore numbers may be expected.

Synoptic links

Try this yourself! Think logically. Write down an important biological term such as 'cell division'. Link related words to it in a 'flow diagram' or 'mind map'. The links will become evident and could form the framework of a synoptic question.

Energy: input and output

This has to be a favourite for many synoptic questions. Energy is involved in so many processes that the frequency of examination will be high.

3 The structure and role of DNA

It is important to know the structure of DNA because it is fundamentally important to the maintenance of life processes and the transfer of characteristics from one generation of a species to the next generation. DNA links into many environmentally and evolutionally based questions.

- The ultimate source of variation is the mutation of DNA. Questions may involve the mechanism of a mutation in terms of DNA change and be followed by natural selection. This can lead to extinction or the formation of a new species. Clearly there are many potential synoptic variations.

- DNA molecules carry the genetic code by which proteins are produced in cells. This links into the production of important proteins. The structure of a protein into primary, secondary, tertiary and quaternary structure may be tested. All enzymes are proteins, so a range of enzymically based question components can be expected in synoptic questions.

- The human genome project is a high profile project. The uses of this human gene 'atlas' will lead to many developments in the coming years. The reporting of developments, radiating from the human genome project, could be the basis of many comprehension type questions, spanning diverse areas of biology courses. Save newspaper cuttings, search the internet, watch documentaries. Note links with genetic diseases, ethics, drugs, etc.

4 Structure and function of the cell surface membrane

There are a range of different mechanisms by which substances can cross the cell surface membrane. These include diffusion, facilitated diffusion, osmosis, active transport, exocytosis and pinocytosis. Additionally glycoproteins have a cell recognition function and some proteins are enzymic in function. Knowledge of these concepts and processes can be tested in cross module questions.

- In an ecologically based question the increasing salinity of a rock pool in sunny conditions could be linked to water potential changes in an aquatic plant or animal. Inter-relationships of organisms within a related food web could follow, identifying such a question as synoptic.

- In cystic fibrosis a transmembrane regulator protein is defective. A mutant gene responsible for the condition codes for a protein with a missing amino acid. This can link to both the correct functioning of the protein, the mechanism of the mutation, and the functioning of the DNA.

5 Transport mechanisms

This theme may unify the following into a synoptic question, transport across membranes, transport mechanisms in animal and plant organs. Additionally they may be linked to homeostatic processes.

- The route of a substance from production in a cell, through a vessel to the consequences of a tissue which receives the substance, could expand into a synoptic question. Homeostasis and negative feedback could well be linked into these ideas.

Sample questions and model answers

A short question which cuts across the course. It refers back to AS. Do not forget those modules! See the *Letts AS Guide* for additional advice and those concepts not found in this volume.

Question 1 *(a short structured question)*

Answer the following questions (a) to (f) in the spaces provided.

(a) Name the kingdom to which all bacteria belong. [1]

Prokaryotae

(b) What is the name given to embryonic plant seed leaves? [1]

cotyledons

(c) In which part of the kidney would you expect to find the glomerulus? [1]

cortex

(d) What is the term used to describe hormone secreting glands? [1]

endocrine

(e) During gametogenesis, name the process which results in the formation of haploid cells? [1]

meiosis

(f) Name the structure which provides the link between nervous and endocrine regulation. [1]

hypothalamus

[Total: 6]

WJEC specimen

Question 2 *(a longer, more open-ended question)*

Living organisms exchange materials with their environment. These exchanges occur across surfaces which have special features. With reference to named examples, discuss how these surfaces are adapted for efficient exchange. [10]

(Quality of written communication assessed in this answer.)

This is an open-ended synoptic question. The examples you would refer to, to illustrate principles of exchange, depend upon which modules you have studied. Alveoli and tracheoles, as well as the structure of cell surface membranes are key aspects in your answer.

- large surface area;
- maintenance of diffusion gradients;
- way(s) in which this is achieved;
- ref. to uptake of organic molecules;
- layer(s) of cells, e.g. squamous epithelium;
- ref. to uptake of ions;
- internal surface of leaf mesophyll;
- channel proteins;
- root hairs; carrier proteins; active uptake;
- any structural detail of surfaces;
- gills/alveoli;
- selectively permeable; permeable/thin;
- ref. to gaseous exchange;
- well ventilated;
- well supplied with blood;
- any detail;
- credit given for other examples such as tracheoles in insects, surface of protoctists, fungi, bacteria – any examples taken from optional modules

Note, there is one mark available for legible text with accurate spelling, punctuation and grammar. [1]

[Total: 10]

OCR specimen

Sample questions and model answers (continued)

Question 3 *(a longer question of higher mark tariff)*

Different concentrations of maltose were placed in the small intestine of a mammal. The amount of glucose appearing in the blood and the small intestine were measured. The results are shown in the graph.

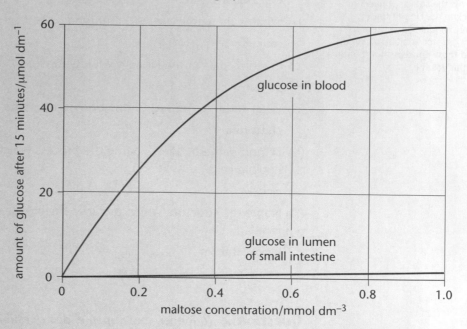

(a) (i) Give the name of the blood vessel most likely to be sampled for glucose. [1]

hepatic portal vein

   (ii) By what chemical process is a molecule of maltose converted into two molecules of glucose? [1]

hydrolysis

(b) The enzyme maltase converts maltose into glucose. This enzyme is found in the cell surface membrane of the epithelial cells of the small intestine.

   (i) Explain the evidence from the graph which supports the view that the breakdown of maltose does not occur in the lumen of the small intestine. [2]

Very little/no increase in the amount of glucose in the lumen;

if breakdown took place in the lumen then it would increase here/

would take some time to diffuse through the wall. [max 2]

   (ii) Suggest an explanation for the shape of the curve showing the change in the amount of glucose in the blood. [3]

More glucose, more active sites occupied;

curve flattens as enzyme becomes limiting;

at any one time, all active sites are occupied. [max 3]

[Total: 7]

Assessment and Qualifications Alliance A specimen

Sample questions and model answers (continued)

Question 4

A cow obtains most of its nutritional requirements from mutualistic microorganisms in its rumen. The diagram summarises the biochemical processes carried out by these microorganisms.

This flow diagram is complex. Follow the input of nutrients and note the reactions which take place. Clearly the microorganisms aid the digestive process of the cow. There are no enzymes shown but you would be aware that enzymes of both the cow and microorganisms would be active in these processes.

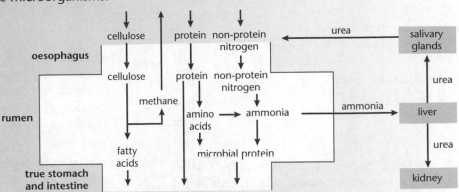

(a) Use the information in the diagram to explain why:

(i) the relation between the cow and the microorganisms which live in its rumen may be described as mutualistic [2]

Both organisms have a nutritional advantage;
cow gains fatty acids/cow gains proteins;
microorganisms gain cellulose/protein/urea. [max 2]

(ii) it is possible for a cow to survive on a diet which is poor in protein [2]

Obtain non-protein nitrogen;
microorganisms convert this to protein;
which cow can digest.

(iii) ruminant animals such as cows are less efficient than non-ruminant animals in converting energy in food into energy in their tissues. [2]

Some food energy is used by microorganisms;
some lost in methane.

(b) Aphids are small insects which feed on plant sap. The table shows the relationship between the amount of soluble nitrogen in plant sap and the body mass and reproductive rate of one species of aphid.

This confirms the synoptic nature of this question! It switches to nitrogenous compounds in plants and nitrogen compound uptake by parasitic aphids.

Soluble nitrogen in plant sap as percentage of dry mass	Mean adult body mass (mg)	Reproductive rate (number of young produced per day per aphid)
2.0	1.8	1.3
2.5	1.6	1.0
3.0	1.3	0.7
3.5	1.0	0.4
4.0	0.7	0.1

Explain how the amount of soluble nitrogen in plant sap affects the body mass and reproductive rate of aphids.

Growth and reproduction requires the production of new tissue;
which has a high protein requirement;
nitrogen is an essential part of protein. [max 2]

[Total: 8]

Assessment and Qualifications Alliance A specimen

Practice examination answers

Chapter 1 Energy for life

1

(a) in cytoplasm [1]

(b) pyruvate [1]

(c) 2 ATPs begin process;
2ATPs are produced from each of the two GP molecules, so −2 + 4 = +2 ATPs net [1]

(d) animal; animal cells produce lactate [1]

(e) oxygen or aerobic [1]

[Total: 5]

2

(a) At this point the amount of carbon dioxide given off by the plant in *respiration*, is totally used by the plant in *photosynthesis*. [2]

(b) compensation point [1]

(c) The continued graph line falls (as light dims); line ends below the horizontal axis (when its dark!). [2]

[Total: 5]

3

(a) $RQ = \dfrac{\text{units of } CO_2}{\text{units of } O_2}$

$0.7 = \dfrac{102}{x}$

$x = \dfrac{102}{0.7}$

$x = 145.7$

$= 146$ [2]

(b) (i) 0.9

(ii) 1.0 [2]

[Total: 4]

4

NaOH absorbs CO_2; as O_2 is taken in, any CO_2 replacing it is absorbed by the NaOH;

liquid in manometer moves;

use the syringe to equalise levels;
volume of oxygen shown by the volume change in the syringe. [max 4]

[Total: 4]

5

(a) Absorption spectrum is obtained from the amount of each wavelength absorbed by the pigments which made up the chlorophyll of the plant.

Action spectrum produced by measuring the amount of photosynthesis by the plant for each separate wavelength. [2]

(b) Low amount of photosynthesis because not much light energy absorbed, most is reflected. [1]

(c) Evolution of oxygen, collected by water displacement. [1]

6

(a) mitochondrion [1]

(b) NADH [1]

(c) cytochrome [1]

(d) ATP [1]

[Total: 4]

7

(a) (i) rate of photosynthesis is proportional to light intensity; rate limited by amount of light available

(ii) as light intensity increases it results in significantly less increase on the rate of photosynthesis

(iii) rate of photosynthesis has levelled off, no longer limited by light (but other conditions could be limiting!). [3]

(b) Similar shape of graph, begins at origin, but graph line above the given plotted curve. [1]

[Total: 4]

Chapter 2 Nutrients

1

(a) cellulose and urea ticked
cellulose and starch ticked
urea and protein ticked
cellulose, starch, urea and protein ticked [4]

(b) bolus is regurgitated;
it is then ground up again by the molars;
(thus increased surface area)
bolus is passed to the omasum then abomasum (true stomach) to be digested further [2]

(c) mutualism (symbiosis would be accepted) [1]
Microorganisms have a habitat at suitable

temperature, microorganisms have a supply of food obtained by cow. [1]
Cow obtains a supply of protein; microorganisms breakdown cellulose into substances which are useful to the cow, e.g. (volatile) fatty acids. [1]

[Total: 9]

2

Fill in the gaps in this order

gastrin; intestinal mucosa; intestinal mucosa; lipids; cholecystokinin. [5]

[Total: 5]

Chapter 3 Control in animals and plants

1

(a) (i) IAA (at these lower) concentrations is *proportional* to the angle of curvature of the stem. [1]

(ii) IAA (at these higher) concentrations is *inversely proportional* to the angle of curvature. [1]

(b) *More* IAA causes the cells at side of stem in contact with agar block to elongate more than other side.

So this side grows more strongly bending stem towards the weaker side. [2]

(c) Growth is only stimulated up to a certain high IAA concentration, after this curvature would be inhibited. [2]

[Total: 6]

2

(a) A = actin
B = myosin [2]

(b) action potential reaches sarcomere [1]

(c) both filaments slide alongside each other;
they form cross bridges;
during contraction the filaments slide together to form a shorter sarcomere [2]

[Total: 5]

3

(i) resting potential achieved; [2]
Na⁺ / K⁺ pump is on

(ii) Na+ / K⁺ pump is off;
so Na⁺ ions enter axon [2]

(iii) maximum depolarisation achieved; K⁺ ions leave [2]

(iv) Na⁺ ions leave due to Na⁺/ K⁺ pump being back on;
this is during the refractory period;
at end of this resting potential re-established;
axon membrane re-polarised [4]

[Total: 10]

Chapter 4 Homeostasis

1 (a)

	Nervous system	Endocrine system
Usually have longer lasting effects		✓
Have cells which secrete transmitter molecules	✓	
Cells communicate by substances in the blood plasma		✓
Use chemicals which bind to receptor sites in cell surface proteins	✓	✓
Involve the use of Na⁺ and K⁺ pumps	✓	

[2]

(b) homeostasis [1]

[Total: 3]

2

It increases permeability of; the collecting ducts, and the distal convoluted tubules of the nephron;
• more water drawn out of the collecting ducts;
• by the sodium and chloride ions;
• in medulla of kidney;
• so more water can be reabsorbed back into blood;
• through capillary network. (max 6) [6]

[Total: 6]

3

(a)

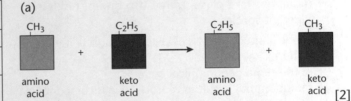

amino acid + keto acid → amino acid + keto acid [2]

(b) (i) liver [1]

(ii) To make different amino acids with the help of the essential amino acids. [2]

[Total: 5]

4

(a) **B**, because as glucose levels rose after meals they did not decrease enough (this kept the blood glucose level too high) [1]

(b) glucose levels fell after every meal, so glucose must have entered the cells and liver [1]

(c) in pancreas;
in the β cells of islets of Langerhans (max 2) [2]

[Total: 4]

Chapter 5 Further genetics

1

(a) no immigration and no emigration; no mutations;
no natural selection; true random mating;
all genotypes must be equally fertile [4]

(b) (i) $q^2 = \dfrac{48}{160}$
= 0.3
q = 0.55

but p + q = 1
so p = 1 – 0.55
= 0.45
but $p^2 + 2pq + q^2 = 1$
so $0.45^2 + 2 \times 0.45 \times 0.55 + 0.55^2 = 1$
0.2 + 0.5 + 0.3 = 1
BB = 0.2 Bb = 0.5 bb = 0.3 [3]

(ii) BB 2000 Bb 5000 bb 3000 [2]

[Total: 9]

2

A (iv), B (iii), C (v), D, (ii), E (i).   [Total: 5]

3

(a) triplet   [1]

(b) codes for an amino acid, codes for stop or start   [2]

(c) Addition CCG ATT CGA TAG <u>CAT</u>
Deletion CCG ATT CGA
Inversion CCG ATT CGA <u>GAT</u>   [3]
[Total: 6]

4

(a) 8 or 4 pairs

(b) (i) During telophase I of meiosis the chromosomes are bivalent/ the centromeres are still intact, whereas in telophase II the chromosomes are single   [1]

(ii) During telophase of mitosis the chromosomes are in pairs, whereas in telophase II of meiosis they are single (haploid)   [2]

(c) the spindle contracts; pulls the centromeres apart; chromosomes begin to be pulled to both poles.   [2]
[Total: 5]

Chapter 6 Biodiversity

1

A = Prokaryotae   C = Protoctista   E = Animalia
B = Fungi   D = Plantae   [5]
[Total: 5]

2

(a) **Allopatric speciation** takes place after geographical isolation;

• the rising of sea level splits a population of animals; formerly connected by land creating two islands;
• mutations take place so that two groups result in different species.

Sympatric speciation takes place through genetic variation;

• in the same geographical area;
• mutation may result in reproductive incompatibility;
• perhaps a structure in birds may lead to a different song being produced by the new variant;
• this may lead to the new variant being rejected from the mainstream group;
• breeding may be possible within its own group of variants.   [6]

(b) Mate them both with a similar male, to give them a chance to produce fertile offspring.

• If they both produce offspring, take a male and female from the offspring, mate them,
• if they produce fertile offspring then original females **are** from the same species.   [2]
[Total: 8]

3

(a)

	mistle-thrush	song-thrush
Kingdom	**Animalia**	**Animalia**
Phylum	Chordata	Chordata
Class	Aves	**Aves**
Order	Passeriformes	Passeriformes
Family	**Turdidae**	**Turdidae**
Genus	**Turdus**	**Turdus**
Species	viscivorus	ericetorum

[3]

(b) disruptive selection   [1]
[Total: 4]

Chapter 7 Ecology and populations

1

(a) no significant migration;
no significant births or deaths;
marking does not have an adverse effect.   [3]

(b) S = total number of individuals in the total population
S_1 = number captured in sample one, marked and released, i.e. 16
S_2 = total number captured in sample two, i.e. 12
S_3 = total marked individuals captured in sample two, i.e. 5

$$\frac{S}{S_1} = \frac{S_2}{S_3} \quad \text{so, } S = \frac{S_1 \times S_2}{S_3}$$

$S = \dfrac{16 \times 12}{5}$ Estimated no. of shrews is 38.4   [2]

(c) Not be very reliable because the numbers are quite low. High population numbers are more reliable.   [1]
[Total: 6]

2

	Type of behaviour			
	kinesis	innate	positive taxis	negative taxis
A bolus of food reaches the top of our oesophagus and is swallowed.		✓		
Insects move from a cold dry area to a warm humid one.			✓	✓
Spring tails (insects) are subjected to increasingly hot conditions, and react by increasing speed in a number of directions. Some go towards the heat source and die.	✓			
A queen bee accepts the advances of a drone bee and is mated.		✓		
A motile alga swims towards light.			✓	

[Total:5]

3
The opposite sexes recognise each other;
the grebes will only mate with other grebes so are more
likely to produce fertile offspring;
mating is synchronised, to coincide with ovulation. [3]
[Total: 3]

Chapter 8 Further effects of human pollution

1
(a) Make sure that water is **homogeneous** when taking
the sample for the colorimeter. [1]
(b) Light is passed through the water sample;
the more particles there are the more light is
absorbed;
so the smaller the absorbance the more the particles
removed by the mussels. [2]
(c) Rate of removal of particles is higher in non-polluted
water, shown by steeper gradient of graph.
Rate of particle removal decreased in polluted water,
shown by the less steep gradient.
So the china clay particles seem to impede removal
of some of the particles;
suggests that the organic molecules cannot be taken
out as efficiently. [3]

(d) increase photosynthesis; more light able to reach
plants because some particles are removed *or*
excreta of mussels decomposed to release mineral ions
e.g. NO_3^- taken in by plants. [1]
[Total: 7]

2
(a) (i) Biochemical oxygen demand [1]
(ii) the amount of **dissolved oxygen** (350g) in a
cubic metres of effluent. Oxygen has depleted
to this level owing to microbial organisms in
the sewage. [2]
(b) B, indicates a greater bacterial presence;
which would add more minerals. [2]
[Total: 5]

Chapter 9 Microbiology

1
(a)

	Virus	Bacterium	Fungus
has membrane-bound organelles			✓
has an outer protein coat of capsomeres	✓		
has ribosomes		✓	✓
is prokaryotic		✓	
is multinucleate			✓
cannot respire	✓		
has plasmids		✓	
has an outer wall of chitin			✓
has nucleic acid but no cytoplasm	✓		

[9]

(b) Bacteriophage is a virus;
which attacks a bacterium;
in order to replicate new viruses. [2]
[Total: 11]

2
(a) (i) CCG AAU UAG CGA UUC AUG mRNA strand [1]
(ii) CCG AAT TAG CGA TTC ATG complementary
DNA strand [1]
(b) reverse transcriptase [1]
[Total: 3]

3
(a) Aseptic transfer of sample from faeces to plate /
use of a sterile loop; incubate at optimum
temperature / 37°C. [2]
(b) Other Gram negative bacteria use lactose in agar;
which results in acid waste;
acid (below pH 6.8) causes the indicator to show these
bacteria as red;
Salmonella bacteria do not use the lactose and remain
colourless. [4]
(c) Bile salts (in MacConkey agar) prevent growth of
Gram positive bacteria. [2]
[Total: 8]

Chapter 10 Biotechnology

1

(a) Steam sterilisation;
microorganisms cannot enter through air filter;
nutrients are pre-sterilised before entry into
fermenter.   [3]

(b) Contaminant microorganisms enter the fermenter;
compete with the *Penicillium*; fungus;
penicillin yield reduced.   [3]

(c) Batch culture: gives best yield; less chance
of contamination.   [2]

[Total: 8]

2

H, A, E, G, C, D, F, B.   [8]

[Total: 8]

3

(a) Beginning of fermentation process shown.
The microorganisms took time to reach
maximum production but kept at this level.
Nutrients constantly added.   [1]

(b) continuous

product amount reaches a constant level;
nutrients at constant level.   [2]

[Total: 3]

Index